Quick Reference
for the
Civil Engineering
PE Exam

Fifth Edition

Michael R. Lindeburg, PE

The Power to Pass
www.ppi2pass.com

Professional Publications, Inc. • Belmont, California

How to Locate and Report Errata for This Book

At PPI, we do our best to bring you error-free books. But when errors do occur, we want to make sure you can view corrections and report any potential errors you find, so the errors cause as little confusion as possible.

A current list of known errata and other updates for this book is available on the PPI website at **www.ppi2pass.com/errata**. We update the errata page as often as necessary, so check in regularly. You will also find instructions for submitting suspected errata. We are grateful to every reader who takes the time to help us improve the quality of our books by pointing out an error.

QUICK REFERENCE FOR THE CIVIL ENGINEERING PE EXAM
Fifth Edition

Current printing of this edition: 2

Printing History

edition number	printing number	update
4	1	Minor corrections. Synchronized with the 10th edition of *Civil Engineering Reference Manual for the PE Exam.*
5	1	Minor corrections. Synchronized with the 11th edition of *Civil Engineering Reference Manual for the PE Exam.*
5	2	Minor corrections.

Printed in the United States of America

PPI
1250 Fifth Avenue, Belmont, CA 94002
(650) 593-9119
www.ppi2pass.com

ISBN: 978-1-59126-138-4

Library of Congress Control Number: 2008928809

Table of Contents

How to Use This Book

This book (and others in the Quick Reference series) was developed to help you minimize problem-solving time on the PE exam. *Quick Reference* is a consolidation of the most useful equations, tables, and figures in the *Civil Engineering Reference Manual*, as well as some bonus equations, tables, and figures to fully equip you for the exam. Using *Quick Reference*, you won't need to wade through pages of descriptive text when all you actually need is a quick look at the formulas to remind you of your next solution step.

The idea is this: you study for the exam using (primarily) your *Reference Manual*, and you take the exam using (primarily) your *Quick Reference*.

Once you've studied and mastered the theory behind an exam topic, you're ready to tackle practice problems. Save time by using this *Quick Reference* right from the start, as you begin solving problems. This book follows the same order and uses the same nomenclature as the *Civil Engineering Reference Manual*. Once you become familiar with the sequencing of subjects in the *Reference Manual*, you'll be at home with *Quick Reference*. As you progress in your understanding of topics, you'll find you can rely more and more on *Quick Reference* for rapid retrieval of equations—without needing to refer back to the *Reference Manual*.

There are also times during problem-solving when you do need access to two kinds of information simultaneously—formulas and data, or formulas and nomenclature, or formulas and theory. We've all experienced the frustration of having to work problems using a spare pencil, a calculator, and the water-bill envelope to serve as page markers in a single book. *Quick Reference* provides a convenient way to keep the equations you need in front of you, even as you may be flip-flopping back and forth between theory and data in your other references.

Quick Reference also provides a convenient place for neatnik engineers to add their own comments and reminders to equations, without having to mess up their primary references. We expect you to throw this book away after the exam, so go ahead and write in it.

Once you start incorporating *Quick Reference* into your problem-solving routine, we predict you won't want to return to the one-book approach. *Quick Reference* will save you precious time—and that's how to use this book.

Codes Used to Prepare This Book

The information that was used to write and update this book was the most current at the time. However, as with engineering practice itself, the PE examination is not always based on the most current codes or cutting-edge technology. Similarly, codes, standards, and regulations adopted by state and local agencies often lag issuance by several years. Thus, the codes that are current, used by you in practice, and tested on the exam can all be different.

PPI lists on its website the dates and editions of the codes, standards, and regulations on which NCEES has announced the PE exams are based. It is your responsibility to find out which codes will be tested on your exam. In the meantime, here are the codes that have been incorporated into this edition.[1]

STRUCTURAL DESIGN STANDARDS

AASHTO: *AASHTO LRFD Bridge Design Specifications*, Third ed., 2004, with 2005 and 2006 Interim Revisions, American Association of State Highway and Transportation Officials, Washington, DC

ACI 318: *Building Code Requirements for Structural Concrete*, 2005, American Concrete Institute, Farmington Hills, MI

ACI 530: *Building Code Requirements for Masonry Structures*, 2005, and ACI 530.1: *Specifications for Masonry Structures*, 2005, American Concrete Institute, Detroit, MI

AISC: *Steel Construction Manual*, Thirteenth ed., 2005, American Institute of Steel Construction, Inc., Chicago, IL

ASCE 7: *Minimum Design Standards for Buildings and Other Structures*, 2005, American Society of Civil Engineers, New York, NY

IBC: *International Building Code*, 2006 ed., International Code Council, Inc., Falls Church, VA

NDS: *National Design Specification for Wood Construction*, 2005 ASD ed., and *National Design Specification*

Supplement, 2005 ASD ed., American Forest and Paper Association, Washington, DC

PCI: *PCI Design Handbook*, Sixth ed., 2004, Precast/Prestressed Concrete Institute, Chicago, IL

TRANSPORTATION DESIGN STANDARDS

AASHTO: *Guide for Design of Pavement Structures*, Fourth ed., 1993, American Association of State Highway and Transportation Officials, Washington, DC[2]

AASHTO: *A Policy on Geometric Design of Highways and Streets*, Fifth ed., 2004, American Association of State Highway and Transportation Officials, Washington, DC

AASHTO: *Roadside Design Guide*, Third ed., 2002, American Association of State Highway and Transportation Officials, Washington, DC

AI: *The Asphalt Handbook* (MS-4), 1989, Asphalt Institute, College Park, MD[2]

HCM: *Highway Capacity Manual (HCM 2000)*, Fourth ed., U.S. Customary version, 2000, Transportation Research Board, National Research Council, Washington, DC

ITE: *Traffic Engineering Handbook*, Fifth ed., 1999, Institute of Transportation Engineers, Washington, DC

MUTCD: *Manual on Uniform Traffic Control Devices*, 2003, U.S. Dept. of Transportation, Federal Highway Administration, Washington, DC

PCA: *Design and Control of Concrete Mixtures*, Fourteenth ed., 2002, Portland Cement Association, Skokie, IL

CONSTRUCTION

ACI 347-04: *Guide to Formwork for Concrete*, 2004, American Concrete Institute, Farmington Hills, MI (in ACI SP-4, Seventh ed. appendix)

ACI SP-4: *Formwork for Concrete*, Seventh ed., 2005, American Concrete Institute, Farmington Hills, MI

---

[1]This is a list of the codes used to prepare *this* book. The *Reference Manual's* introduction contains a complete list of the references that you should consider bringing to the exam.

[2]Though these codes are not specified by NCEES, they are integral to the information presented in Chs. 75 and 76 of the *Civil Engineering Reference Manual*.

AISC: *Steel Construction Manual*, Thirteenth ed., American Institute of Steel Construction, Inc., Chicago, IL

ASCE 37-02: *Design Loads on Structures During Construction*, 2002, American Society of Civil Engineers, Reston, VA

CMWB: *Standard Practice for Bracing Masonry Walls During Construction*, 2001, Council for Masonry Wall Bracing, Mason Contractors Association of America, Lombard, IL

MUTCD-Pt 6: *Manual on Uniform Traffic Control Devices—Part 6 Temporary Traffic Control*, 2003, U.S. Federal Highway Administration

NDS: *National Design Specification for Wood Construction*, 2005, American Forest & Paper Association/American Wood Council, Washington, DC

OSHA: *Occupational Safety and Health Standards for the Construction Industry*, 29 CFR Part 1926, (U.S. federal version), U.S. Department of Labor, Washington, DC

For Instant Recall

Fundamental and Physical Constants

quantity	symbol	English	SI
Density			
air [STP] [32°F (0°C)]		0.0805 lbm/ft^3	1.29 kg/m^3
air [70°F (20°C), 1 atm]		0.0749 lbm/ft^3	1.20 kg/m^3
earth [mean]		345 lbm/ft^3	5520 kg/m^3
mercury		849 lbm/ft^3	1.360×10^4 kg/m^3
seawater		64.0 lbm/ft^3	1025 kg/m^3
water [mean]		62.4 lbm/ft^3	1000 kg/m^3
Specific Gravity			
mercury			13.6
water			1.0
Gravitational Acceleration			
earth [mean]	g	32.174 (32.2) ft/sec^2	9.8067 (9.81) m/s^2
Pressure, atmospheric		14.696 (14.7) lbf/in^2	1.0133×10^5 Pa
Temperature, standard		32°F (492°R)	0°C (273K)
Fundamental Constants			
Avogadro's number	N_A		6.022×10^{23} mol^{-1}
gravitational constant	g_c	32.174 lbm-ft/lbf-sec^2	
specific gas constant, air	R	53.35 ft-lbf/lbm-°R	287.03 J/kg·K
specific gas constant, methane	R_{CH_4}	96.32 ft-lbf/lbm-°R	518.3 J/kg·K
universal gas constant	R^*	1545.4 ft-lbf/lbmol-°R	8314.570 J/kmol·K
	R^*	1.986 BTU/lbmol-°R	0.08206 atm·L/mol·K
Molecular Weight			
air			29
carbon			12
carbon dioxide			44
helium			4
hydrogen			2
methane			16
nitrogen			28
oxygen			32
Steel			
modulus of elasticity		2.9×10^7 psi	
modulus of shear		1.2×10^7 psi	
Poisson's ratio		0.3	

Temperature Conversions

$$°F = 32 + \tfrac{9}{5}°C$$

$$°C = \tfrac{5}{9}(°F - 32)$$

$$°R = °F + 460$$

$$K = °C + 273$$

$$\Delta°R = \tfrac{9}{5}\Delta K$$

$$\Delta K = \tfrac{5}{9}\Delta°R$$

SI Prefixes

symbol	prefix	value
a	atto	10^{-18}
f	femto	10^{-15}
p	pico	10^{-12}
n	nano	10^{-9}
μ	micro	10^{-6}
m	milli	10^{-3}
c	centi	10^{-2}
d	deci	10^{-1}
da	deka	10
h	hecto	10^2
k	kilo	10^3
M	mega	10^6
G	giga	10^9
T	tera	10^{12}
P	peta	10^{15}
E	exa	10^{18}

Engineering Conversions

Unless noted otherwise, atmospheres are standard; Btus are IT (international table); calories are gram-calories; gallons are U.S. liquid; miles are statute; pounds-mass are avoirdupois; chains are surveyors'.

multiply	by	to obtain
ac	10.0	chain^2
ac	43,560	ft^2
ac	0.40469	hectare
ac	4046.87	m^2
ac	1/640	mi^2
ac-ft	43,560	ft^3
ac-ft	0.01875	mi^2-in
ac-ft	1233.5	m^3
ac-ft	325,851	gal
ac-ft/day	0.50416	ft^3/s
ac-ft/mi^2	0.01875	in (runoff)
ac-in	102,790	L
ac-in/hr	1.0083	ft^3/s
angstrom	1.0×10^{-10}	m
atm	1.01325	bar
atm	76.0	cm Hg
atm	33.90	ft water
atm	29.921	in Hg
atm	14.696	lbf/in^2
atm	101.33	kPa
atm	1.0133×10^5	Pa
bar	0.9869	atm
bar	1.0×10^5	Pa
Btu	778.169	ft-lbf
Btu	1055.056	J
Btu	2.928×10^{-4}	kW-hr
Btu	1.0×10^{-5}	therm
Btu/hr	0.21611	ft-lbf/s
Btu/hr	3.929×10^{-4}	hp
Btu/hr	0.29307	W
Btu/lbm	2.3260	kJ/kg
Btu/lbm-°R	4.1868	kJ/kg·K
Btu/s	778.26	ft-lbf/s
Btu/s	46,680	ft-lbf/min
Btu/s	1.4148	hp
Btu/s	1.0545	kW
cal	3.968×10^{-3}	Btu
cal	4.1868	J
chain	66.0	ft
chain	1/80	mi (statute)
chain	4.0	rod
chain	22	yd
chain^2	1/10	ac
cm	0.03281	ft
cm	0.39370	in
cm/s	1.9686	ft/min
cm^2	0.1550	in^2
cm^3	2.6417×10^{-4}	gal
cm^3	0.0010	L
day (mean solar)	86,400	s
day (sidereal)	86,164.09	s
darcy	1.0623×10^{-11}	ft^2
degree (angular)	1/0.9	grad
degree (angular)	$2\pi/360$	radian
degree (angular)	17.778	mil
dyne	1.0×10^{-5}	N
eV	1.6022×10^{-19}	J
ft	1/66	chain
ft	0.3048	m
ft	1/5280	mi
ft	0.06060	rod
ft of water	0.43328	lbf/in^2
ft-kips	1.358	kN·m
ft-lbf (energy)	1.2851×10^{-3}	Btu
ft-lbf (energy)	1.3558	J
ft-lbf (energy)	3.766×10^{-7}	kW·h
ft-lbf (torque)	1.3558	N·m
ft-lbf/min	1/46,680	Btu/s
ft-lbf/min	1/33,000	hp
ft-lbf/min	2.2598×10^{-5}	kW

multiply	by	to obtain
ft-lbf/s	1.2849×10^{-3}	Btu/s
ft-lbf/s	1/550	hp
ft-lbf/s	1.3558×10^{-3}	kW
ft/min	0.50798	cm/s
ft/s	0.68180	mi/hr
ft^2	1/43,560	ac
ft^2	9.4135×10^{10}	darcy
ft^2	0.09290	m^2
ft^3	1/43,560	ac-ft
ft^3	1728.0	in^3
ft^3	7.4805	gal
ft^3	28.320	L
ft^3	0.03704	yd^3
ft^3/day	0.005195	gal/min
ft^3/mi^2	4.3×10^{-7}	in runoff
ft^3/min	10,772	gal/day
ft^3/s	1.9835	ac-ft/day
ft^3/s	0.99177	ac-in/hr
ft^3/s	646,317	gal/day
ft^3/s	448.83	gal/min
ft^3/s	0.03719	mi^2-in/day
ft^3/s	0.64632	MGD
g	0.03527	oz
g	0.002205	lbm
g/cm^3	1000.0	kg/m^3
g/cm^3	62.428	lbm/ft^3
gal	1/325,851	ac-ft
gal	3785.4	cm^3
gal	0.13368	ft^3
gal (Imperial)	1.2	gal (U.S.)
gal (U.S.)	0.8327	gal (Imperial)
gal	3.7854	L
gal	3.7854×10^{-3}	m^3
gal	1.0×10^{-6}	MG
gal	0.00495	yd^3
gal (of water)	8.34	lbm (of water)
gal/ac-day	0.04356	MG/ft^2-day
gal/day	9.283×10^{-5}	ft^3/min
gal/day	1.5472×10^{-6}	ft^3/s
gal/day	1/1440	gal/min
gal/day	1.0×10^{-6}	MG/day
gal/day-ft	0.01242	m^3/m·day
gal/day-ft^2	0.04075	m^3/m^2·day
gal/day-ft^2	1.0	Meinzer unit
gal/day-ft^2	0.04355	MGD/ac (mgad)
gal/min	0.002228	ft^3/s
gal/min	192.50	ft^3/day
gal/min	1440	gal/day
gal/min	0.06309	L/s
grad	0.90	degrees (angular)
grain	1.4286×10^{-4}	lbm
grain/gal	142.86	lbm/MG
grain/gal	17.118	ppm
grain/gal	17.118	mg/L
hectare	2.4711	ac
hectare	10,000	m^2
hp	550.0	ft-lbf/s
hp	33,000	ft-lbf/min
hp	0.7457	kW
hp	2545	Btu/hr
hp	0.70678	Btu/s
hp-hr	2545.2	Btu
in	2.540	cm
in	0.02540	m
in	25.40	mm
in Hg	0.4910	lbf/in^2
in Hg	70.704	lbf/ft^2
in Hg	13.60	in water
in runoff	53.30	ac-ft/mi^2
in runoff	2.3230×10^6	ft^3/mi^2
in water	5.1990	lbf/ft^2
in water	0.0361	lbf/in^2
in water	0.07353	in Hg
in-lbf	0.11298	N·m
in/ft	1/0.012	mm/m
in^2	6.4516	cm^2
in^3	1/1728	ft^3
J	9.4778×10^{-4}	Btu
J	6.2415×10^{18}	eV
J	0.73756	ft-lbf

multiply	by	to obtain
J	1.0	N·m
J/s	1.0	W
kg	2.2046	lbm
kg/m³	0.06243	lbm/ft³
kip	4.4480	kN
kip	1000.0	lbf
kip	4448.0	N
kip/ft	14.594	kN/m
kip/ft²	47.880	kPa
kJ	0.94778	Btu
kJ	737.56	ft-lbf
kJ/kg	0.42992	Btu/lbm
kJ/kg·K	0.23885	Btu/lbm-°R
km	3280.8	ft
km	0.62138	mi
km/hr	0.62138	mi/hr
kN	0.2248	kips
kN·m	0.73757	ft-kips
kN/m	0.06852	kips/ft
kPa	9.8692×10^{-3}	atm
kPa	0.14504	lbf/in²
kPa	1000.0	Pa
kPa	0.02089	kips/ft²
ksi	6.8947×10^{6}	Pa
kW	737.56	ft-lbf/s
kW	44,250	ft-lbf/min
kW	1.3410	hp
kW	3413.0	Btu/hr
kW	0.9483	Btu/s
kW-hr	3413.0	Btu
kW-hr	3.60×10^{6}	J
L	1/102,790	ac-in
L	1000.0	cm³
L	0.03531	ft³
L	0.26417	gal
L	61.024	in³
L	0.0010	m³
L/s	2.1189	ft³/min
L/s	15.850	gal/min
lbf	0.001	kips
lbf	4.4482	N
lbf/ft²	0.01414	in Hg
lbf/ft²	0.19234	in water
lbf/ft²	0.00694	lbf/in²
lbf/ft²	47.880	Pa
lbf/ft²	5.0×10^{-4}	tons/ft²
lbf/in²	0.06805	atm
lbf/in²	144.0	lbf/ft²
lbf/in²	2.3080	ft water
lbf/in²	27.70	in water
lbf/in²	2.0370	in Hg
lbf/in²	6894.8	Pa
lbf/in²	0.00050	tons/in²
lbf/in²	0.0720	tons/ft²
lbm	7000.0	grains
lbm	453.59	g
lbm	0.45359	kg
lbm	4.5359×10^{5}	mg
lbm	5.0×10^{-4}	tons (mass)
lbm (of water)	0.12	gal (of water)
lbm/ac-ft-day	0.02296	lbm/1000 ft³-day
lbm/ft³	0.016018	g/cm³
lbm/ft³	16.018	kg/m³
lbm/1000 ft³-day	43.560	lbm/ac-ft-day
lbm/1000 ft³-day	133.68	lbm/MG-day
lbm/MG	0.0070	grains/gal
lbm/MG	0.11983	mg/L
lbm/MG-day	0.00748	lbm/1000 ft³-day
leagues	4428.0	m
m	1.0×10^{10}	angstroms
m	3.2808	ft
m	39.370	in
m	2.2583×10^{-4}	leagues
m	1.0936	yd
m/s	196.85	ft/min
m²	2.4711×10^{-4}	ac
m²	10.764	ft²
m²	1/10,000	hectare
mi²-in	53.3	ac-ft

multiply	by	to obtain
mi²-in/day	26.89	ft³/s
m³	8.1071×10^{-4}	ac-ft
m³/m·day	80.5196	gal/day-ft
m³/m²·day	24.542	gal/day-ft²
Meinzer unit	1.0	gal/day-ft²
mg	2.2046×10^{-6}	lbm
mg/L	1.0	ppm
mg/L	0.05842	grains/gal
mg/L	8.3454	lbm/MG
MG	1.0×10^{6}	gal
MG/ac-day	22.968	gal/ft²-day
MGD	1.5472	ft³/sec
MGD	1×10^{6}	gal/day
MGD/ac (mgad)	22.957	gal/day-ft²
mi	5280.0	ft
mi	80.0	chains
mi	1.6093	km
mi (statute)	0.86839	miles (nautical)
mi	320.0	rods
mi/hr	1.4667	ft/s
mi²	640.0	acres
micron	1.0×10^{-6}	m
micron	0.001	mm
mil (angular)	0.05625	degrees
mil (angular)	3.375	min
min (angular)	0.29630	mils
min (angular)	2.90888×10^{-4}	radians
min (time, mean solar)	60	s
mm	1/25.4	in
mm	1000.0	microns
mm/m	0.012	in/ft
MPa	1.0×10^{6}	Pa
N	0.22481	lbf
N	1.0×10^{5}	dynes
N·m	0.73756	ft-lbf
N·m	8.8511	in-lbf
N·m	1.0	J
N/m²	1.0	Pa
oz	28.353	g
Pa	0.001	kPa
Pa	1.4504×10^{-7}	ksi
Pa	1.4504×10^{-4}	lbf/in²
Pa	0.02089	lbf/ft²
Pa	1.0×10^{-6}	MPa
Pa	1.0	N/m²
ppm	0.05842	grains/gal
radian	$180/\pi$	degrees (angular)
radian	3437.7	min (angular)
rod	0.250	chain
rod	16.50	ft
rod	1/320	mi
s (time)	1/86,400	day (mean solar)
s (time)	1.1605×10^{-5}	day (sidereal)
s (time)	1/60	min
therm	1.0×10^{5}	Btu
ton (force)	2000.0	lbf
ton (mass)	2000.0	lbm
ton/ft²	2000.0	lbf/ft²
ton/ft²	13.889	lbf/in²
W	3.413	Btu/hr
W	0.73756	ft-lbf/s
W	1.3410×10^{-3}	hp
W	1.0	J/s
yd	1/22	chain
yd	0.91440	m
yd³	27	ft³
yd³	201.97	gal

Mensuration

Triangle

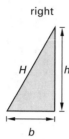 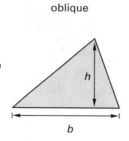 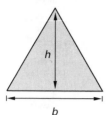

equilateral   right   oblique

$$A = \tfrac{1}{2}bh = \frac{\sqrt{3}}{4}b^2 \qquad A = \tfrac{1}{2}bh \qquad A = \tfrac{1}{2}bh$$

$$h = \frac{\sqrt{3}}{2}b \qquad\qquad H^2 = b^2 + h^2$$

Circle

$$p = 2\pi r$$

$$A = \pi r^2 = \frac{p^2}{4\pi}$$

Circular Segment

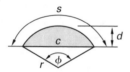

$$A = \tfrac{1}{2}r^2(\phi - \sin\phi)$$

$$\phi = \frac{s}{r} = 2\left(\arccos\frac{r-d}{r}\right)$$

$$c = 2r\sin\left(\frac{\phi}{2}\right)$$

Circular Sector

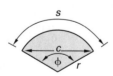

$$A = \tfrac{1}{2}\phi r^2 = \tfrac{1}{2}sr$$

$$\phi = \frac{s}{r}$$

$$s = r\phi$$

$$c = 2r\sin\left(\frac{\phi}{2}\right)$$

Parabola

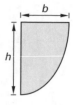

$$A = \tfrac{2}{3}bh$$

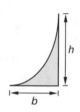

$$A = \tfrac{1}{3}bh$$

Ellipse

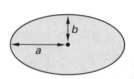

$$A = \pi ab$$

$$p \approx 2\pi\sqrt{\tfrac{1}{2}(a^2 + b^2)} \qquad \left[\begin{array}{c}\text{Euler's}\\ \text{upper bound}\end{array}\right]$$

Trapezoid

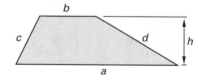

$$p = a + b + c + d$$
$$A = \tfrac{1}{2}h(a + b)$$

The trapezoid is isosceles if $c = d$.

Regular Polygon (n equal sides)

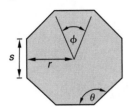

$$\phi = \frac{2\pi}{n}$$

$$\theta = \frac{\pi(n-2)}{n} = \pi - \phi$$

$$p = ns$$

$$s = 2r \tan\left(\frac{\theta}{2}\right)$$

$$A = \tfrac{1}{2}nsr$$

Sphere

$$V = \frac{4\pi r^3}{3}$$

$$A = 4\pi r^2$$

Right Circular Cone

$$V = \frac{\pi r^2 h}{3}$$

$$A = \pi r \sqrt{r^2 + h^2}$$

(does not include base area)

Right Circular Cylinder

$$V = \pi r^2 h$$

$$A = 2\pi r h$$

(does not include end area)

Spherical Segment (Spherical Cap)

Surface area of a spherical segment of radius r cut out by an angle θ_0 rotated from the center about a radius, r, is

$$A = 2\pi r^2 \left(1 - \cos\theta_0\right)$$

$$\omega = \frac{A}{r^2} = 2\pi \left(1 - \cos\theta_0\right)$$

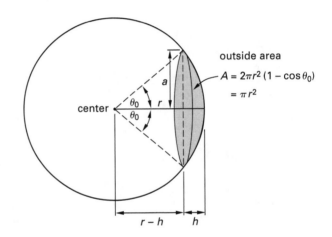

Background and Support

CERM Chapter 3
Algebra

Chapter, section, equation, figure, and table numbers correspond to CERM. For additional study material, go to the corresponding chapter and section number in CERM.

9. ROOTS OF QUADRATIC EQUATIONS

$$x_1, x_2 = \frac{-b \pm \sqrt{b^2 - 4ac}}{2a} \qquad 3.17$$

13. RULES FOR EXPONENTS AND RADICALS

$$(ab)^n = a^n b^n \qquad 3.24$$

$$b^{m/n} = \sqrt[n]{b^m} = \left(\sqrt[n]{b}\right)^m \qquad 3.25$$

$$(b^n)^m = b^{nm} \qquad 3.26$$

$$b^m b^n = b^{m+n} \qquad 3.27$$

15. LOGARITHM IDENTITIES

$$\log_b(b) = 1 \qquad 3.34$$

$$\log_b(1) = 0 \qquad 3.35$$

$$\log_b(b^n) = n \qquad 3.36$$

$$\log(x^a) = a \log(x) \qquad 3.37$$

$$\log(xy) = \log(x) + \log(y) \qquad 3.41$$

$$\log\left(\frac{x}{y}\right) = \log(x) - \log(y) \qquad 3.42$$

$$\log_a(x) = \log_b(x) \log_a(b) \qquad 3.43$$

$$\ln(x) = \log_{10}(x) \ln(10) \approx 2.3026 \log_{10}(x) \qquad 3.44$$

$$\log_{10}(x) = \ln(x) \log_{10}(e) \approx 0.4343 \ln(x) \qquad 3.45$$

CERM Chapter 4
Linear Algebra

Chapter, section, equation, figure, and table numbers correspond to CERM. For additional study material, go to the corresponding chapter and section number in CERM.

5. DETERMINANTS

$$\mathbf{A} = \begin{bmatrix} a & b \\ c & d \end{bmatrix}$$

$$|\mathbf{A}| = \begin{vmatrix} a & b \\ c & d \end{vmatrix} = ad - bc \qquad 4.3$$

$$\mathbf{A} = \begin{bmatrix} a & b & c \\ d & e & f \\ g & h & i \end{bmatrix}$$

$$|\mathbf{A}| = a \begin{vmatrix} e & f \\ h & i \end{vmatrix} - d \begin{vmatrix} b & c \\ h & i \end{vmatrix} + g \begin{vmatrix} b & c \\ e & f \end{vmatrix} \qquad 4.6$$

CERM Chapter 5
Vectors

Chapter, section, equation, figure, and table numbers correspond to CERM. For additional study material, go to the corresponding chapter and section number in CERM.

2. VECTORS IN *n*-SPACE

$$|\mathbf{V}| = \sqrt{(x_2 - x_1)^2 + (y_2 - y_1)^2} \qquad 5.1$$

$$\phi = \arctan\left(\frac{y_2 - y_1}{x_2 - x_1}\right) \qquad 5.2$$

Figure 5.1 *Vector in Two-Dimensional Space*

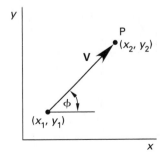

3. UNIT VECTORS

$$\mathbf{V} = |\mathbf{V}|\mathbf{a} = V_x\mathbf{i} + V_y\mathbf{j} + V_z\mathbf{k} \qquad 5.8$$

Figure 5.3 *Cartesian Unit Vectors*

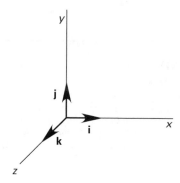

8. VECTOR DOT PRODUCT

Figure 5.5 *Vector Dot Product*

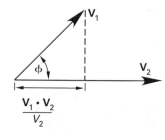

$$\mathbf{V}_1 \cdot \mathbf{V}_2 = |\mathbf{V}_1||\mathbf{V}_2|\cos\phi$$
$$= V_{1x}V_{2x} + V_{1y}V_{2y} + V_{1z}V_{2z} \qquad 5.21$$

9. VECTOR CROSS PRODUCT

$$|\mathbf{V}_1 \times \mathbf{V}_2| = |\mathbf{V}_1||\mathbf{V}_2|\sin\phi \qquad 5.32$$

Figure 5.6 *Vector Cross Product*

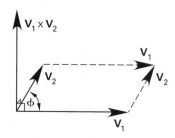

CERM Chapter 6
Trigonometry

> Chapter, section, equation, figure, and table numbers correspond to CERM. For additional study material, go to the corresponding chapter and section number in CERM.

1. DEGREES AND RADIANS

multiply	by	to obtain
radians	$\dfrac{180}{\pi}$	degrees
degrees	$\dfrac{\pi}{180}$	radians

4. RIGHT TRIANGLES

$$x^2 + y^2 = r^2 \qquad 6.1$$

5. CIRCULAR TRANSCENDENTAL FUNCTIONS

$$\text{sine: } \sin\theta = \frac{y}{r} = \frac{\text{opposite}}{\text{hypotenuse}} \qquad 6.2$$

$$\text{cosine: } \cos\theta = \frac{x}{r} = \frac{\text{adjacent}}{\text{hypotenuse}} \qquad 6.3$$

$$\text{tangent: } \tan\theta = \frac{y}{x} = \frac{\text{opposite}}{\text{adjacent}} \qquad 6.4$$

$$\text{cotangent: } \cot\theta = \frac{x}{y} = \frac{\text{adjacent}}{\text{opposite}} \qquad 6.5$$

$$\text{secant: } \sec\theta = \frac{r}{x} = \frac{\text{hypotenuse}}{\text{adjacent}} \qquad 6.6$$

$$\text{cosecant: } \csc\theta = \frac{r}{y} = \frac{\text{hypotenuse}}{\text{opposite}} \qquad 6.7$$

$$\cot\theta = \frac{1}{\tan\theta} \qquad 6.8$$

$$\sec\theta = \frac{1}{\cos\theta} \qquad 6.9$$

$$\csc\theta = \frac{1}{\sin\theta} \qquad 6.10$$

Figure 6.6 *Trigonometric Functions in a Unit Circle*

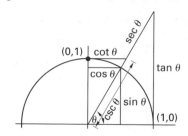

6. SMALL ANGLE APPROXIMATIONS

$$\sin\theta \approx \tan\theta \approx \theta\Big|_{\theta<10° \ (0.175 \text{ rad})} \qquad 6.11$$

$$\cos\theta \approx 1\Big|_{\theta<5° \ (0.0873 \text{ rad})} \qquad 6.12$$

10. TRIGONOMETRIC IDENTITIES

$$\sin^2\theta + \cos^2\theta = 1 \qquad 6.14$$
$$1 + \tan^2\theta = \sec^2\theta \qquad 6.15$$
$$1 + \cot^2\theta = \csc^2\theta \qquad 6.16$$

- *double-angle formulas*:

$$\sin 2\theta = 2\sin\theta\cos\theta = \frac{2\tan\theta}{1+\tan^2\theta} \qquad 6.17$$

$$\cos 2\theta = \cos^2\theta - \sin^2\theta = 1 - 2\sin^2\theta$$
$$= 2\cos^2\theta - 1 = \frac{1-\tan^2\theta}{1+\tan^2\theta} \qquad 6.18$$

$$\tan 2\theta = \frac{2\tan\theta}{1-\tan^2\theta} \qquad 6.19$$

$$\cot 2\theta = \frac{\cot^2\theta - 1}{2\cot\theta} \qquad 6.20$$

- *two-angle formulas*:

$$\sin(\theta \pm \phi) = \sin\theta\cos\phi \pm \cos\theta\sin\phi \qquad 6.21$$

$$\cos(\theta \pm \phi) = \cos\theta\cos\phi \mp \sin\theta\sin\phi \qquad 6.22$$

$$\tan(\theta \pm \phi) = \frac{\tan\theta \pm \tan\phi}{1 \mp \tan\theta\tan\phi} \qquad 6.23$$

$$\cot(\theta \pm \phi) = \frac{\cot\phi\cot\theta \mp 1}{\cot\phi \pm \cot\theta} \qquad 6.24$$

- *half-angle formulas* $(\theta < 180°)$:

$$\sin\frac{\theta}{2} = \sqrt{\frac{1-\cos\theta}{2}} \qquad 6.25$$

$$\cos\frac{\theta}{2} = \sqrt{\frac{1+\cos\theta}{2}} \qquad 6.26$$

$$\tan\frac{\theta}{2} = \sqrt{\frac{1-\cos\theta}{1+\cos\theta}} = \frac{\sin\theta}{1+\cos\theta} = \frac{1-\cos\theta}{\sin\theta} \qquad 6.27$$

- *miscellaneous formulas* $(\theta < 90°)$:

$$\sin\theta = 2\sin\left(\frac{\theta}{2}\right)\cos\left(\frac{\theta}{2}\right) \qquad 6.28$$

$$\sin\theta = \sqrt{\frac{1-\cos 2\theta}{2}} \qquad 6.29$$

$$\cos\theta = \cos^2\left(\frac{\theta}{2}\right) - \sin^2\left(\frac{\theta}{2}\right) \qquad 6.30$$

$$\cos\theta = \sqrt{\frac{1+\cos 2\theta}{2}} \qquad 6.31$$

$$\tan\theta = \frac{2\tan\left(\frac{\theta}{2}\right)}{1-\tan^2\left(\frac{\theta}{2}\right)}$$
$$= \frac{2\sin\left(\frac{\theta}{2}\right)\cos\left(\frac{\theta}{2}\right)}{\cos^2\left(\frac{\theta}{2}\right) - \sin^2\left(\frac{\theta}{2}\right)} \qquad 6.32$$

$$\tan\theta = \sqrt{\frac{1-\cos 2\theta}{1+\cos 2\theta}}$$
$$= \frac{\sin 2\theta}{1+\cos 2\theta} = \frac{1-\cos 2\theta}{\sin 2\theta} \qquad 6.33$$

$$\cot\theta = \frac{\cot^2\left(\frac{\theta}{2}\right) - 1}{2\cot\left(\frac{\theta}{2}\right)}$$
$$= \frac{\cos^2\left(\frac{\theta}{2}\right) - \sin^2\left(\frac{\theta}{2}\right)}{2\sin\left(\frac{\theta}{2}\right)\cos\left(\frac{\theta}{2}\right)} \qquad 6.34$$

$$\cot\theta = \sqrt{\frac{1+\cos 2\theta}{1-\cos 2\theta}}$$
$$= \frac{1+\cos 2\theta}{\sin 2\theta} = \frac{\sin 2\theta}{1-\cos 2\theta} \qquad 6.35$$

14. GENERAL TRIANGLES

$$\frac{\sin A}{a} = \frac{\sin B}{b} = \frac{\sin C}{c} \qquad 6.57$$

$$a^2 = b^2 + c^2 - 2bc\cos A \qquad 6.58$$

Figure 6.9 *General Triangle*

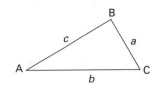

CERM Chapter 7
Analytic Geometry

Chapter, section, equation, figure, and table numbers correspond to CERM. For additional study material, go to the corresponding chapter and section number in CERM.

10. STRAIGHT LINES

- *general form:*
$$Ax + By + C = 0 \qquad 7.8$$
$$A = -mB \qquad 7.9$$
$$B = \frac{-C}{b} \qquad 7.10$$
$$C = -aA = -bB \qquad 7.11$$

- *slope-intercept form:*
$$y = mx + b \qquad 7.12$$
$$m = \frac{-A}{B} = \tan\theta = \frac{y_2 - y_1}{x_2 - x_1} \qquad 7.13$$
$$b = \frac{-C}{B} \qquad 7.14$$
$$a = \frac{-C}{A} \qquad 7.15$$

- *point-slope form:*
$$y - y_1 = m(x - x_1) \qquad 7.16$$

- *intercept form:*
$$\frac{x}{a} + \frac{y}{b} = 1 \qquad 7.17$$

- *two-point form:*
$$\frac{y - y_1}{x - x_1} = \frac{y_2 - y_1}{x_2 - x_1} \qquad 7.18$$

Figure 7.8 *Straight Line*

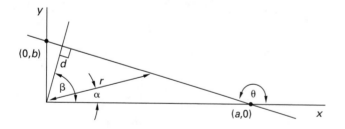

14. DISTANCES BETWEEN GEOMETRIC FIGURES

- between two points in (x, y, z) format:
$$d = \sqrt{(x_2 - x_1)^2 + (y_2 - y_1)^2 + (z_2 - z_1)^2} \qquad 7.48$$

- between a point (x_0, y_0) and a line $Ax + By + C = 0$:
$$d = \frac{|Ax_0 + By_0 + C|}{\sqrt{A^2 + B^2}} \qquad 7.49$$

- between a point (x_0, y_0, z_0) and a plane $Ax + By + Cz + D = 0$:
$$d = \frac{|Ax_0 + By_0 + Cz_0 + D|}{\sqrt{A^2 + B^2 + C^2}} \qquad 7.50$$

- between two parallel lines $Ax + By + C = 0$:
$$d = \left| \frac{|C_2|}{\sqrt{A_2^2 + B_2^2}} - \frac{|C_1|}{\sqrt{A_1^2 + B_1^2}} \right| \qquad 7.51$$

17. CIRCLE

$$Ax^2 + Ay^2 + Dx + Ey + F = 0 \qquad 7.66$$
$$(x - h)^2 + (y - k)^2 = r^2 \qquad 7.67$$

18. PARABOLA

$$(y - k)^2 = 4p(x - h) \Big|_{\text{opens horizontally}} \qquad 7.73$$
$$y^2 = 4px \Big|_{\substack{\text{vertex at origin} \\ h=k=0}} \qquad 7.74$$

Figure 7.13 *Parabola*

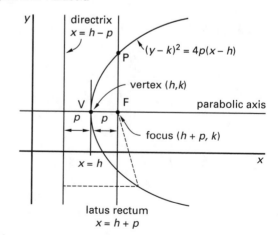

CERM Chapter 11
Probability and Statistical Analysis of Data

Chapter, section, equation, figure, and table numbers correspond to CERM. For additional study material, go to the corresponding chapter and section number in CERM.

12. POISSON DISTRIBUTION

$$p\{x\} = f(x) = \frac{e^{-\lambda}\lambda^x}{x!} \qquad [\lambda > 0] \qquad 11.34$$

λ is both the mean and the variance of the Poisson distribution.

15. NORMAL DISTRIBUTION

$$z = \frac{x_0 - \mu}{\sigma} \quad \text{[standard normal variable]} \qquad 11.43$$

16. APPLICATION: RELIABILITY

$$R\{t\} = e^{-\lambda t} = e^{-t/\text{MTTF}} \qquad 11.47$$

$$\lambda = \frac{1}{\text{MTTF}} \qquad 11.48$$

$$R\{t\} = 1 - F(t) = 1 - (1 - e^{-\lambda t}) = e^{-\lambda t} \qquad 11.49$$

$$z\{t\} = \lambda \qquad 11.50$$

18. MEASURES OF CENTRAL TENDENCY

$$\overline{x} = \left(\frac{1}{n}\right)(x_1 + x_2 + \cdots + x_n) = \frac{\sum x_i}{n} \qquad 11.56$$

$$\text{geometric mean} = \sqrt[n]{x_1 x_2 x_3 \cdots x_n} \quad [x_i > 0] \qquad 11.57$$

$$\text{harmonic mean} = \frac{n}{\dfrac{1}{x_1} + \dfrac{1}{x_2} + \cdots + \dfrac{1}{x_n}} \qquad 11.58$$

$$\text{root mean square} = x_{\text{rms}} = \sqrt{\frac{\sum x_i^2}{n}} \qquad 11.59$$

19. MEASURES OF DISPERSION

$$\sigma = \sqrt{\frac{\sum(x_i - \mu)^2}{N}} = \sqrt{\frac{\sum x_i^2}{N} - \mu^2} \qquad 11.60$$

$$s = \sqrt{\frac{\sum(x_i - \overline{x})^2}{n - 1}} = \sqrt{\frac{\sum x_i^2 - \dfrac{(\sum x_i)^2}{n}}{n - 1}} \qquad 11.61$$

$$\sigma_{\text{sample}} = s\sqrt{\frac{n - 1}{n}} \qquad 11.62$$

$$\text{coefficient of variation} = \frac{s}{\overline{x}} \qquad 11.63$$

22. APPLICATION: CONFIDENCE LIMITS

$$\text{LCL:} \quad \mu - z_c\sigma$$
$$\text{UCL:} \quad \mu + z_c\sigma$$

Table 11.1 *Values of z for Various Confidence Levels*

confidence level, C	one-tail limit z	two-tail limit z
90%	1.28	1.645
95%	1.645	1.96
97.5%	1.96	2.17
99%	2.33	2.575
99.5%	2.575	2.81
99.75%	2.81	3.00

Areas Under the Standard Normal Curve
(0 to z)

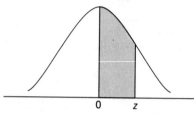

z	0	1	2	3	4	5	6	7	8	9
0.0	0.0000	0.0040	0.0080	0.0120	0.0160	0.0199	0.0239	0.0279	0.0319	0.0359
0.1	0.0398	0.0438	0.0478	0.0517	0.0557	0.0596	0.0636	0.0675	0.0714	0.0754
0.2	0.0793	0.0832	0.0871	0.0910	0.0948	0.0987	0.1026	0.1064	0.1103	0.1141
0.3	0.1179	0.1217	0.1255	0.1293	0.1331	0.1368	0.1406	0.1443	0.1480	0.1517
0.4	0.1554	0.1591	0.1628	0.1664	0.1700	0.1736	0.1772	0.1808	0.1844	0.1879
0.5	0.1915	0.1950	0.1985	0.2019	0.2054	0.2088	0.2123	0.2157	0.2190	0.2224
0.6	0.2258	0.2291	0.2324	0.2357	0.2389	0.2422	0.2454	0.2486	0.2518	0.2549
0.7	0.2580	0.2612	0.2642	0.2673	0.2704	0.2734	0.2764	0.2794	0.2823	0.2852
0.8	0.2881	0.2910	0.2939	0.2967	0.2996	0.3023	0.3051	0.3078	0.3106	0.3133
0.9	0.3159	0.3186	0.3212	0.3238	0.3264	0.3289	0.3315	0.3340	0.3365	0.3389
1.0	0.3413	0.3438	0.3461	0.3485	0.3508	0.3531	0.3554	0.3577	0.3599	0.3621
1.1	0.3643	0.3665	0.3686	0.3708	0.3729	0.3749	0.3770	0.3790	0.3810	0.3830
1.2	0.3849	0.3869	0.3888	0.3907	0.3925	0.3944	0.3962	0.3980	0.3997	0.4015
1.3	0.4032	0.4049	0.4066	0.4082	0.4099	0.4115	0.4131	0.4147	0.4162	0.4177
1.4	0.4192	0.4207	0.4222	0.4236	0.4251	0.4265	0.4279	0.4292	0.4306	0.4319
1.5	0.4332	0.4345	0.4357	0.4370	0.4382	0.4394	0.4406	0.4418	0.4429	0.4441
1.6	0.4452	0.4463	0.4474	0.4484	0.4495	0.4505	0.4515	0.4525	0.4535	0.4545
1.7	0.4554	0.4564	0.4573	0.4582	0.4591	0.4599	0.4608	0.4616	0.4625	0.4633
1.8	0.4641	0.4649	0.4656	0.4664	0.4671	0.4678	0.4686	0.4693	0.4699	0.4706
1.9	0.4713	0.4719	0.4726	0.4732	0.4738	0.4744	0.4750	0.4756	0.4761	0.4767
2.0	0.4772	0.4778	0.4783	0.4788	0.4793	0.4798	0.4803	0.4808	0.4812	0.4817
2.1	0.4821	0.4826	0.4830	0.4834	0.4838	0.4842	0.4846	0.4850	0.4854	0.4857
2.2	0.4861	0.4864	0.4868	0.4871	0.4875	0.4878	0.4881	0.4884	0.4887	0.4890
2.3	0.4893	0.4896	0.4898	0.4901	0.4904	0.4906	0.4909	0.4911	0.4913	0.4916
2.4	0.4918	0.4920	0.4922	0.4925	0.4927	0.4929	0.4931	0.4932	0.4934	0.4936
2.5	0.4938	0.4940	0.4941	0.4943	0.4945	0.4946	0.4948	0.4949	0.4951	0.4952
2.6	0.4953	0.4955	0.4956	0.4957	0.4959	0.4960	0.4961	0.4962	0.4963	0.4964
2.7	0.4965	0.4966	0.4967	0.4968	0.4969	0.4970	0.4971	0.4972	0.4973	0.4974
2.8	0.4974	0.4975	0.4976	0.4977	0.4977	0.4978	0.4979	0.4979	0.4980	0.4981
2.9	0.4981	0.4982	0.4982	0.4983	0.4984	0.4984	0.4985	0.4985	0.4986	0.4986
3.0	0.4987	0.4987	0.4987	0.4988	0.4988	0.4989	0.4989	0.4989	0.4990	0.4990
3.1	0.4990	0.4991	0.4991	0.4991	0.4992	0.4992	0.4992	0.4992	0.4993	0.4993
3.2	0.4993	0.4993	0.4994	0.4994	0.4994	0.4994	0.4994	0.4995	0.4995	0.4995
3.3	0.4995	0.4995	0.4996	0.4996	0.4996	0.4996	0.4996	0.4996	0.4996	0.4997
3.4	0.4997	0.4997	0.4997	0.4997	0.4997	0.4997	0.4997	0.4997	0.4997	0.4998
3.5	0.4998	0.4998	0.4998	0.4998	0.4998	0.4998	0.4998	0.4998	0.4998	0.4998
3.6	0.4998	0.4998	0.4999	0.4999	0.4999	0.4999	0.4999	0.4999	0.4999	0.4999
3.7	0.4999	0.4999	0.4999	0.4999	0.4999	0.4999	0.4999	0.4999	0.4999	0.4999
3.8	0.4999	0.4999	0.4999	0.4999	0.4999	0.4999	0.4999	0.4999	0.4999	0.4999
3.9	0.5000	0.5000	0.5000	0.5000	0.5000	0.5000	0.5000	0.5000	0.5000	0.5000

CERM Chapter 13
Energy, Work, and Power

> Chapter, section, equation, figure, and table numbers correspond to CERM. For additional study material, go to the corresponding chapter and section number in CERM.

3. WORK

$$W_{\text{variable force}} = \int \mathbf{F} \cdot d\mathbf{s} \quad \text{[linear systems]} \qquad 13.4$$

$$W_{\text{variable torque}} = \int \mathbf{T} \cdot d\boldsymbol{\theta} \quad \text{[rotational systems]} \qquad 13.5$$

$$W_{\text{constant force}} = \mathbf{F} \cdot \mathbf{s} = Fs\cos\phi \quad \text{[linear systems]} \qquad 13.6$$

$$W_{\text{constant torque}} = \mathbf{T} \cdot \boldsymbol{\theta} = Fr\theta\cos\phi$$
$$\text{[rotational systems]} \qquad 13.7$$

$$W_{\text{friction}} = F_f s \qquad 13.8$$

$$W_{\text{gravity}} = \left(\frac{mg}{g_c}\right)(h_2 - h_1) \qquad 13.9(b)$$

$$W_{\text{spring}} = \tfrac{1}{2}k\left(\delta_2^2 - \delta_1^2\right) \qquad 13.10$$

4. POTENTIAL ENERGY OF A MASS

$$E_{\text{potential}} = \frac{mgh}{g_c} \qquad 13.11(b)$$

5. KINETIC ENERGY OF A MASS

$$E_{\text{kinetic}} = \frac{mv^2}{2g_c} \qquad 13.13(b)$$

$$E_{\text{rotational}} = \frac{I\omega^2}{2g_c} \qquad 13.15(b)$$

6. SPRING ENERGY

$$E_{\text{spring}} = \tfrac{1}{2}k\delta^2 \qquad 13.16$$

8. INTERNAL ENERGY OF A MASS

$$Q = mc\,\Delta T \qquad 13.19$$

9. WORK-ENERGY PRINCIPLE

$$W = \Delta E = E_2 - E_1 \qquad 13.24$$

11. POWER

$$P = \frac{W}{\Delta t} \qquad 13.25$$

$$P = Fv \quad \text{[linear systems]} \qquad 13.26$$

$$P = T\omega \quad \text{[rotational systems]} \qquad 13.27$$

$$P = \dot{m}\Delta u \quad \text{[flowing fluid]} \qquad 13.28$$

Water Resources

CERM Chapter 14
Fluid Properties

Chapter, section, equation, figure, and table numbers correspond to CERM. For additional study material, go to the corresponding chapter and section number in CERM.

4. DENSITY

$$\rho = \frac{p}{RT} \quad \text{[gases]} \qquad 14.5$$

5. SPECIFIC VOLUME

$$v = \frac{1}{\rho} \qquad 14.6$$

6. SPECIFIC GRAVITY

$$\text{SG}_{\text{liquid}} = \frac{\rho_{\text{liquid}}}{\rho_{\text{water}}} \qquad 14.7$$

$$\text{SG}_{\text{gas}} = \frac{\rho_{\text{gas}}}{\rho_{\text{air}}} \qquad 14.8$$

$$\text{SG}_{\text{gas}} = \frac{\text{MW}_{\text{gas}}}{\text{MW}_{\text{air}}} = \frac{\text{MW}_{\text{gas}}}{29.0}$$

$$= \frac{R_{\text{air}}}{R_{\text{gas}}} = \frac{53.3 \, \dfrac{\text{ft-lbf}}{\text{lbm-}°\text{R}}}{R_{\text{gas}}} \qquad 14.9$$

7. SPECIFIC WEIGHT

$$\gamma = \rho \times \frac{g}{g_c} \qquad 14.14(b)$$

9. VISCOSITY

$$\tau = \mu \frac{d\text{v}}{dy} \qquad 14.18$$

The constant of proportionality, μ, is the absolute viscosity.

10. KINEMATIC VISCOSITY

$$\nu = \frac{\mu g_c}{\rho} \qquad 14.19(b)$$

18. BULK MODULUS

The bulk modulus, E, is the reciprocal of compressibility, β.

$$E = \frac{1}{\beta} \qquad 14.36$$

19. SPEED OF SOUND

$$a = \sqrt{\frac{Eg_c}{\rho}} = \sqrt{\frac{g_c}{\beta\rho}} \quad \text{[liquids]} \qquad 14.37(b)$$

$$a = \sqrt{kg_cRT} = \sqrt{\frac{kg_cR^*T}{\text{MW}}} \quad \text{[gases]} \qquad 14.38(b)$$

CERM Chapter 15
Fluid Statics

Chapter, section, equation, figure, and table numbers correspond to CERM. For additional study material, go to the corresponding chapter and section number in CERM.

2. MANOMETERS

$$p_2 - p_1 = \rho_m \times \frac{g}{g_c} \times h$$

$$= \gamma_m h \qquad 15.4$$

$$p_2 - p_1 = \frac{g}{g_c} \times (\rho_m h + \rho_1 h_1 - \rho_2 h_2)$$

$$= \gamma_m h + \gamma_1 h_1 - \gamma_2 h_2 \qquad 15.5(b)$$

Figure 15.4 Manometer Requiring Corrections

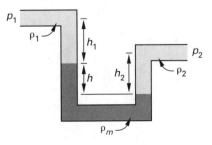

3. HYDROSTATIC PRESSURE

$$p = \frac{F}{A} \qquad 15.6$$

4. FLUID HEIGHT EQUIVALENT TO PRESSURE

$$p = \frac{\rho g h}{g_c} = \gamma h \qquad 15.7(b)$$

7. PRESSURE ON A RECTANGULAR VERTICAL PLANE SURFACE

Figure 15.9 *Hydrostatic Pressure on a Vertical Plane Surface*

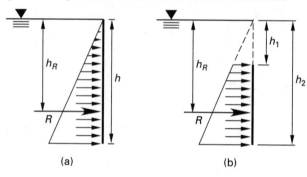

(a)   (b)

$$\overline{p} = \frac{\frac{1}{2}\rho g (h_1 + h_2)}{g_c} = \frac{1}{2}\gamma(h_1 + h_2) \qquad 15.15(b)$$

$$R = \overline{p}A \qquad 15.16$$

$$h_R = \frac{2}{3}\left(h_1 + h_2 - \frac{h_1 h_2}{h_1 + h_2}\right) \qquad 15.17$$

9. PRESSURE ON A GENERAL PLANE SURFACE

$$\overline{p} = \frac{\rho g h_c \,\sin\theta}{g_c} = \gamma h_c \sin\theta \qquad 15.24(b)$$

$$R = \overline{p}A \qquad 15.25$$

$$h_R = h_c + \frac{I_c}{A\,h_c} \qquad 15.26$$

Figure 15.11 *General Plane Surface*

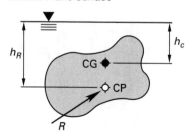

13. HYDROSTATIC FORCES ON A DAM

$$M_{\text{overturning}} = R_x \times y_{R_x} \qquad 15.31$$

$$M_{\text{resisting}} = (R_y \times x_{R_y}) + (W \times x_{\text{CG}}) \qquad 15.32$$

$$(\text{FS})_{\text{overturning}} = \frac{M_{\text{resisting}}}{M_{\text{overturning}}} \qquad 15.33$$

$$F_f = \mu_{\text{static}}\ N = \mu_{\text{static}}(W + R_y) \qquad 15.34$$

$$(\text{FS})_{\text{sliding}} = \frac{F_f}{R_x} \qquad 15.35$$

$$p_{\max}, p_{\min} = \left(\frac{R_y + W}{b}\right)\left(1 \pm \frac{6e}{b}\right)$$
$$[\text{per unit width}] \qquad 15.36$$

$$e = \frac{b}{2} - x_{\text{v}} \qquad 15.37$$

$$x_{\text{v}} = \frac{M_{\text{resisting}}}{R_y + W} \qquad 15.38$$

Figure 15.16 *Dam*

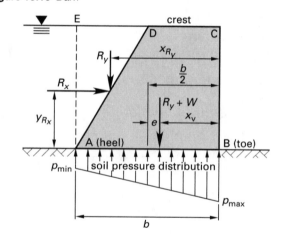

18. BUOYANCY

$$F_{\text{buoyant}} = \frac{\rho g V_{\text{displaced}}}{g_c} = \gamma V_{\text{displaced}} \qquad 15.56(b)$$

$$\text{SG} = \frac{W_{\text{dry}}}{W_{\text{dry}} - W_{\text{submerged}}} \qquad 15.57$$

CERM Chapter 16
Fluid Flow Parameters

Chapter, section, equation, figure, and table numbers correspond to CERM. For additional study material, go to the corresponding chapter and section number in CERM.

2. KINETIC ENERGY

$$E_v = \frac{v^2}{2g_c} \qquad \text{16.3(b)}$$

3. POTENTIAL ENERGY

$$E_z = \frac{zg}{g_c} \qquad \text{16.5(b)}$$

4. PRESSURE ENERGY

$$E_p = \frac{p}{\rho} \qquad \text{16.6}$$

5. BERNOULLI EQUATION

$$E_t = \frac{p}{\rho} + \frac{v^2}{2g_c} + \frac{zg}{g_c} \qquad \text{16.11(b)}$$

8. HYDRAULIC RADIUS

$$r_h = \frac{\text{area in flow}}{\text{wetted perimeter}} = \frac{A}{s} \qquad \text{16.18}$$

9. EQUIVALENT DIAMETER

$$D_e = 4r_h \qquad \text{16.20}$$

Table 16.1 Equivalent Diameters for Common Conduit Shapes

conduit cross section	D_e
flowing full	
circle	D
annulus (outer diameter D_o, inner diameter D_i)	$D_o - D_i$
square (side L)	L
rectangle (sides L_1 and L_2)	$\dfrac{2L_1 L_2}{L_1 + L_2}$
flowing partially full	
half-filled circle (diameter D)	D
rectangle (h deep, L wide)	$\dfrac{4hL}{L + 2h}$
wide, shallow stream (h deep)	$4h$
triangle (h deep, L broad, s side)	$\dfrac{hL}{s}$
trapezoid (h deep, a wide at top, b wide at bottom, s side)	$\dfrac{2h(a + b)}{b + 2s}$

10. REYNOLDS NUMBER

$$\text{Re} = \frac{D_e v \rho}{g_c \mu} \qquad \text{16.22(b)}$$

$$\text{Re} = \frac{D_e v}{\nu} \qquad \text{16.23}$$

16. SPECIFIC ENERGY

$$E_{\text{specific}} = E_p + E_v \qquad \text{16.36}$$

$$E_{\text{specific}} = \frac{p}{\rho} + \frac{v^2}{2g_c} \qquad \text{16.37(b)}$$

CERM Chapter 17
Fluid Dynamics

Chapter, section, equation, figure, and table numbers correspond to CERM. For additional study material, go to the corresponding chapter and section number in CERM.

9. ENERGY LOSS DUE TO FRICTION: TURBULENT FLOW

- *Darcy equation*

$$h_f = \frac{f L \mathrm{v}^2}{2Dg} \qquad 17.28$$

- *Hazen-Williams equation*

$$h_{f,\text{feet}} = \frac{3.022 \mathrm{v}_{\text{ft/sec}}^{1.85} L_{\text{ft}}}{C^{1.85} D^{1.17}} \qquad 17.30$$

$$h_{f,\text{feet}} = \frac{10.44 L_{\text{ft}} \dot{V}_{\text{gpm}}^{1.85}}{C^{1.85} d_{\text{inches}}^{4.87}} \qquad 17.31$$

15. MINOR LOSSES

$$h_m = K h_{\mathrm{v}} \qquad 17.41$$

$$K = \frac{f L_e}{D} \qquad 17.42$$

- *sudden enlargements:* (D_1 is the smaller of the two diameters)

$$K = \left(1 - \left(\frac{D_1}{D_2}\right)^2\right)^2 \qquad 17.43$$

- *sudden contractions:* (D_1 is the smaller of the two diameters)

$$K = \tfrac{1}{2}\left(1 - \left(\frac{D_1}{D_2}\right)^2\right) \qquad 17.44$$

16. VALVE FLOW COEFFICIENTS

$$Q_{\text{gpm}} = C_{\mathrm{v}} \sqrt{\frac{\Delta p_{\text{psi}}}{\text{SG}}} \qquad 17.50$$

21. DISCHARGE FROM TANKS

$$\mathrm{v}_t = \sqrt{2gh} \qquad 17.67$$

$$h = z_1 - z_2 \qquad 17.68$$

$$\mathrm{v}_o = C_{\mathrm{v}} \sqrt{2gh} \qquad 17.69$$

$$C_{\mathrm{v}} = \frac{\text{actual velocity}}{\text{theoretical velocity}} = \frac{\mathrm{v}_o}{\mathrm{v}_t} \qquad 17.70$$

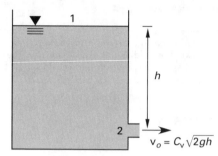
Figure 17.10 *Discharge from a Tank*

$$C_c = \frac{\text{area of vena contracta}}{\text{orifice area}} \qquad 17.74$$

$$C_d = C_{\mathrm{v}} C_c$$
$$= \frac{\text{actual discharge}}{\text{theoretical discharge}} \qquad 17.76$$

23. COORDINATES OF A FLUID STREAM

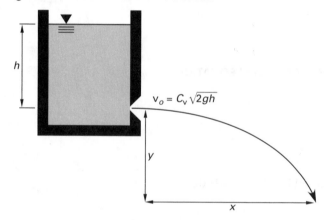
Figure 17.13 *Coordinates of a Fluid Stream*

$$\mathrm{v}_x = \mathrm{v}_o \qquad 17.78$$

$$x = \mathrm{v}_o t = \mathrm{v}_o \sqrt{\frac{2y}{g}} = 2C_{\mathrm{v}}\sqrt{hy} \qquad 17.79$$

$$\mathrm{v}_y = gt \qquad 17.80$$

$$y = \frac{gt^2}{2} = \frac{gx^2}{2\mathrm{v}_o^2} = \frac{x^2}{4hC_{\mathrm{v}}^2} \qquad 17.81$$

33. PITOT-STATIC GAUGE

$$\frac{v^2}{2g_c} = \frac{p_t - p_s}{\rho} = \frac{h(\rho_m - \rho)}{\rho} \times \left(\frac{g}{g_c}\right) \qquad \textit{17.144(b)}$$

$$v = \sqrt{\frac{2gh(\rho_m - \rho)}{\rho}} \qquad \textit{17.145}$$

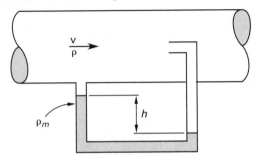

Figure 17.23 *Pitot-Static Gauge*

34. VENTURI METER

$$v_2 = \left(\frac{C_v}{\sqrt{1 - \left(\frac{A_2}{A_1}\right)^2}}\right)\sqrt{\frac{2g_c(p_1 - p_2)}{\rho}} \qquad \textit{17.148(b)}$$

$$\beta = \frac{D_2}{D_1} \qquad \textit{17.149}$$

$$F_{va} = \frac{1}{\sqrt{1 - \left(\frac{A_2}{A_1}\right)^2}} = \frac{1}{\sqrt{1 - \beta^4}} \qquad \textit{17.150}$$

$$v_2 = \left(\frac{C_v}{\sqrt{1 - \beta^4}}\right)\sqrt{\frac{2g(\rho_m - \rho)h}{\rho}}$$

$$= C_v F_{va}\sqrt{\frac{2g(\rho_m - \rho)h}{\rho}} \qquad \textit{17.151}$$

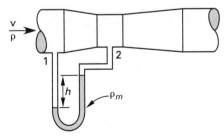

Figure 17.25 *Venturi Meter with Manometer*

$$\dot{V} = C_d A_2 v_{2,\text{ideal}} \qquad \textit{17.152}$$

$$C_f = C_d F_{va} = \frac{C_d}{\sqrt{1 - \beta^4}} \qquad \textit{17.153}$$

$$\dot{V} = C_f A_2 \sqrt{\frac{2g(\rho_m - \rho)h}{\rho}} \qquad \textit{17.154}$$

35. ORIFICE METER

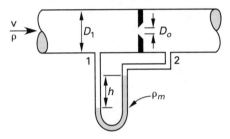

Figure 17.27 *Orifice Meter with Differential Manometer*

$$A_2 = C_c A_o \qquad \textit{17.155}$$

$$v_o = \left(\frac{C_v}{\sqrt{1 - \left(\frac{C_c A_o}{A_1}\right)^2}}\right)\sqrt{\frac{2g_c(p_1 - p_2)}{\rho}} \qquad \textit{17.156(b)}$$

$$v_o = \left(\frac{C_v}{\sqrt{1 - \left(\frac{C_c A_o}{A_1}\right)^2}}\right)\sqrt{\frac{2g(\rho_m - \rho)h}{\rho}} \qquad \textit{17.157}$$

$$F_{va} = \frac{1}{\sqrt{1 - \left(\frac{C_c A_o}{A_1}\right)^2}} \qquad \textit{17.158}$$

$$C_f = C_d F_{va} \qquad \textit{17.159}$$

$$\dot{V} = C_f A_o \sqrt{\frac{2g(\rho_m - \rho)h}{\rho}} = C_f A_o \sqrt{\frac{2g_c(p_1 - p_2)}{\rho}}$$

$$\textit{17.160(b)}$$

47. CONFINED STREAMS IN PIPE BENDS

$$F_x = p_2 A_2 \cos\theta - p_1 A_1 + \frac{\dot{m}(v_2 \cos\theta - v_1)}{g_c}$$

17.200(b)

$$F_y = \left(p_2 A_2 + \frac{\dot{m} v_2}{g_c}\right) \sin\theta$$

17.201(b)

Figure 17.40 Pipe Bend

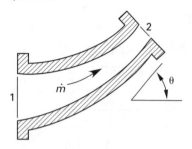

48. WATER HAMMER

$$t = \frac{L}{a} \quad \text{[one way]}$$

17.202

$$t = \frac{2L}{a} \quad \text{[round trip]}$$

17.203

$$\Delta p = \frac{\rho a \Delta v}{g_c}$$

17.204(b)

Figure 17.41 Water Hammer

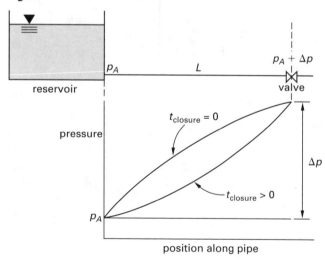

52. DRAG

$$F_D = \frac{C_D A \rho v^2}{2 g_c}$$

17.212(b)

54. TERMINAL VELOCITY

$$v = \sqrt{\frac{2(mg - F_b)}{C_D A \rho_{\text{fluid}}}} = \sqrt{\frac{2 V g(\rho_{\text{object}} - \rho_{\text{fluid}})}{C_D A \rho_{\text{fluid}}}}$$

17.216

$$v = \sqrt{\frac{4 D g(\rho_{\text{sphere}} - \rho_{\text{fluid}})}{3 C_D \rho_{\text{fluid}}}} \quad \text{[sphere]}$$

17.217

$$v = \frac{D^2(\rho_{\text{sphere}} - \rho_{\text{fluid}})}{18 \mu} \times \left(\frac{g}{g_c}\right)$$

$$\begin{bmatrix} \text{small particles;} \\ \text{Stokes' law} \end{bmatrix}$$

17.218(b)

58. SIMILARITY

$$L_r = \frac{\text{size of model}}{\text{size of prototype}}$$

17.223

59. VISCOUS AND INERTIAL FORCES DOMINATE

$$\text{Re}_m = \text{Re}_p$$

17.226

$$\frac{L_m v_m}{\nu_m} = \frac{L_p v_p}{\nu_p}$$

17.227

60. INERTIAL AND GRAVITATIONAL FORCES DOMINATE

$$\text{Fr} = \frac{v^2}{Lg}$$

17.228

$$\text{Re}_m = \text{Re}_p$$

17.229

$$\text{Fr}_m = \text{Fr}_p$$

17.230

$$\frac{\nu_m}{\nu_p} = \left(\frac{L_m}{L_p}\right)^{3/2} = (L_r)^{3/2}$$

17.231

$$n_r = (L_r)^{1/6}$$

17.232

61. SURFACE TENSION FORCE DOMINATES

$$\text{We} = \frac{v^2 L \rho}{\sigma}$$

17.233

$$\text{We}_m = \text{We}_p$$

17.234

CERM Chapter 18
Hydraulic Machines

> Chapter, section, equation, figure, and table numbers correspond to CERM. For additional study material, go to the corresponding chapter and section number in CERM.

6. TYPES OF CENTRIFUGAL PUMPS

$$v_{tip} = \frac{\pi D n}{60 \frac{\text{sec}}{\text{min}}} = \frac{D\omega}{2} \qquad 18.3$$

10. PUMPING POWER

Table 18.5 Hydraulic Horsepower Equations[a]

	Q (gal/min)	$\dot{m}$ (lbm/sec)	$\dot{V}$ (ft³/sec)
h_A in feet	$\dfrac{h_A Q(SG)}{3956}$	$\left(\dfrac{h_A \dot{m}}{550}\right) \times \left(\dfrac{g}{g_c}\right)$	$\dfrac{h_A \dot{V}(SG)}{8.814}$
Δp in psi[b]	$\dfrac{\Delta p Q}{1714}$	$\left(\dfrac{\Delta p \dot{m}}{(238.3)(SG)}\right) \times \left(\dfrac{g}{g_c}\right)$	$\dfrac{\Delta p \dot{V}}{3.819}$
Δp in psf[b]	$\dfrac{\Delta p Q}{2.468 \times 10^5}$	$\left(\dfrac{\Delta p \dot{m}}{(34,320)(SG)}\right) \times \left(\dfrac{g}{g_c}\right)$	$\dfrac{\Delta p \dot{V}}{550}$
W in $\dfrac{\text{ft-lbf}}{\text{lbm}}$	$\dfrac{W Q(SG)}{3956}$	$\dfrac{W \dot{m}}{550}$	$\dfrac{W \dot{V}(SG)}{8.814}$

(Multiply horsepower by 0.7457 to obtain kilowatts.)
[a]Table 18.5 is based on $\rho_{water} = 62.4$ lbm/ft³ and $g = 32.2$ ft/sec².
[b]Velocity head changes must be included in Δp.

Table 18.6 Hydraulic Kilowatt Equations[a]

	Q (L/s)	$\dot{m}$ (kg/s)	$\dot{V}$ (m³/s)
h_A in meters	$\dfrac{(9.81)h_A Q(SG)}{1000}$	$\dfrac{(9.81)h_A \dot{m}}{1000}$	$(9.81)h_A \dot{V}(SG)$
Δp in kPa[b]	$\dfrac{\Delta p Q}{1000}$	$\dfrac{\Delta p \dot{m}}{1000(SG)}$	$\Delta p \dot{V}$
W in $\dfrac{\text{J}}{\text{kg}}$[b]	$\dfrac{W Q(SG)}{1000}$	$\dfrac{W \dot{m}}{1000}$	$W \dot{V}(SG)$

(Multiply kilowatts by 1.341 to obtain horsepower.)
[a]Table 18.6 is based on $\rho_{water} = 1000$ kg/m³ and $g = 9.81$ m/s².
[b]Velocity head changes must be included in Δp.

11. PUMPING EFFICIENCY

$$\eta = \eta_p \eta_m = \frac{\text{hydraulic horsepower}}{\text{motor horsepower}} \qquad 18.16$$

12. COST OF ELECTRICITY

$$\text{cost} = \frac{W_{kW\text{-}hr} \times \text{cost per kW-hr}}{\eta_m} \qquad 18.18$$

13. STANDARD MOTOR SIZES AND SPEEDS

$$n = \frac{120 \times f}{\text{no. of poles}} \quad \text{[synchronous speed]} \qquad 18.19$$

$$\begin{array}{l}\text{slip} \\ \text{(in rpm)}\end{array} = \text{synchronous speed} - \text{actual speed} \qquad 18.20$$

$$\begin{array}{l}\text{percent} \\ \text{slip}\end{array} = 100\% \times \frac{\begin{array}{c}\text{synchronous speed} \\ - \text{actual speed}\end{array}}{\text{synchronous speed}} \qquad 18.21$$

$$\text{kVA rating} = \frac{\text{motor power in kW}}{\text{power factor}} \qquad 18.22$$

14. PUMP SHAFT LOADING

$$T_{in\text{-}lbf} = \frac{63{,}025 \times P_{hp}}{n} \qquad 18.23$$

$$T_{ft\text{-}lbf} = \frac{5252 \times P_{hp}}{n} \qquad 18.24$$

$$\text{overhung load} = \frac{2KT}{D_{sheave}} \qquad 18.27$$

15. SPECIFIC SPEED

$$n_s = \frac{n\sqrt{Q}}{(h_A)^{0.75}} \qquad 18.28(b)$$

17. NET POSITIVE SUCTION HEAD

$$\frac{NPSHR_2}{NPSHR_1} = \left(\frac{Q_2}{Q_1}\right)^2 \qquad 18.29$$

$$NPSHA = h_{atm} + h_{z(s)} - h_{f(s)} - h_{vp} - h_{ac} \qquad 18.31$$

$$NPSHA = h_{p(s)} + h_{v(s)} - h_{vp} - h_{ac} \qquad 18.32$$

$$NPSHA < NPSHR \quad \text{[criterion for cavitation]} \qquad 18.33$$

20. SUCTION SPECIFIC SPEED

$$n_{ss} = \frac{n\sqrt{Q}}{(NPSHR \text{ in ft})^{0.75}} \qquad 18.36(b)$$

22. SYSTEM CURVES

$$\frac{h_{f,1}}{h_{f,2}} = \left(\frac{Q_1}{Q_2}\right)^2 \qquad 18.40$$

26. AFFINITY LAWS

(Impeller size is constant; speed is varied.)

$$\frac{Q_2}{Q_1} = \frac{n_2}{n_1} \qquad 18.41$$

$$\frac{h_2}{h_1} = \left(\frac{n_2}{n_1}\right)^2 = \left(\frac{Q_2}{Q_1}\right)^2 \qquad 18.42$$

$$\frac{P_2}{P_1} = \left(\frac{n_2}{n_1}\right)^3 = \left(\frac{Q_2}{Q_1}\right)^3 \qquad 18.43$$

$$\frac{Q_2}{Q_1} = \frac{D_2}{D_1} \qquad 18.44$$

$$\frac{h_2}{h_1} = \left(\frac{D_2}{D_1}\right)^2 \qquad 18.45$$

$$\frac{P_2}{P_1} = \left(\frac{D_2}{D_1}\right)^3 \qquad 18.46$$

27. PUMP SIMILARITY

(Impeller size varies.)

$$\frac{n_1 D_1}{\sqrt{h_1}} = \frac{n_2 D_2}{\sqrt{h_2}} \qquad 18.48$$

$$\frac{Q_1}{D_1^2 \sqrt{h_1}} = \frac{Q_2}{D_2^2 \sqrt{h_2}} \qquad 18.49$$

$$\frac{P_1}{\rho_1 D_1^2 h_1^{1.5}} = \frac{P_2}{\rho_2 D_2^2 h_2^{1.5}} \qquad 18.50$$

$$\frac{Q_1}{n_1 D_1^3} = \frac{Q_2}{n_2 D_2^3} \qquad 18.51$$

$$\frac{P_1}{\rho_1 n_1^3 D_1^5} = \frac{P_2}{\rho_2 n_2^3 D_2^5} \qquad 18.52$$

$$\frac{n_1 \sqrt{Q_1}}{h_1^{0.75}} = \frac{n_2 \sqrt{Q_2}}{h_2^{0.75}} \qquad 18.53$$

CERM Chapter 19
Open Channel Flow

Chapter, section, equation, figure, and table numbers correspond to CERM. For additional study material, go to the corresponding chapter and section number in CERM.

5. PARAMETERS USED IN OPEN CHANNEL FLOW

- *hydraulic radius*

$$R = \frac{A}{P} \qquad 19.2$$

- *hydraulic depth*

$$D_h = \frac{A}{w} \qquad 19.3$$

- *energy gradient* (geometric slope)

$$S_0 = \frac{\Delta z}{L} \quad \text{[uniform flow]} \qquad 19.7$$

6. GOVERNING EQUATIONS FOR UNIFORM FLOW

$$n = \text{Manning constant}$$

$$C = \left(\frac{1.49}{n}\right) R^{1/6} \quad \text{[Chezy equation]} \qquad 19.11(b)$$

$$\text{v} = \left(\frac{1.49}{n}\right) R^{2/3} \sqrt{S} \quad \left[\begin{matrix} \text{Chezy-Manning} \\ \text{equation} \end{matrix}\right] \quad 19.12(b)$$

$$Q = \text{v}A = \left(\frac{1.49}{n}\right) AR^{2/3} \sqrt{S}$$

$$= K\sqrt{S} \qquad 19.13(b)$$

8. HAZEN-WILLIAMS VELOCITY

$$\text{v} = 1.318 C R^{0.63} S_0^{0.54} \qquad 19.14(b)$$

9. NORMAL DEPTH

$$d_n = 0.788 \left(\frac{nQ}{w\sqrt{S}}\right)^{3/5} \quad [w \gg d_n] \qquad 19.15(b)$$

$$D = d_n = 1.335 \left(\frac{nQ}{\sqrt{S}}\right)^{3/8} \quad \text{[full]} \qquad 19.16(b)$$

$$D = 2d_n = 1.731 \left(\frac{nQ}{\sqrt{S}}\right)^{3/8} \quad \text{[half full]}$$
$$19.17(b)$$

$$R = \frac{wd_n}{w + 2d_n} \qquad 19.18$$

$$A = wd_n \qquad 19.19$$

$$Q = \left(\frac{1.49}{n}\right)(wd_n)\left(\frac{wd_n}{w + 2d_n}\right)^{2/3} \sqrt{S} \quad \text{[rectangular]}$$
$$19.20(b)$$

11. SIZING TRAPEZOIDAL AND RECTANGULAR CHANNELS

Figure 19.2 *Trapezoidal Cross Section*

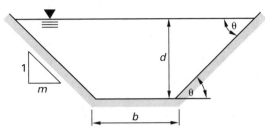

$$Q = \frac{K'b^{8/3}\sqrt{S_0}}{n} \qquad 19.31$$

$$K' = \left(\frac{1.49\left(1 + m\left(\frac{d}{b}\right)\right)^{5/3}}{\left(1 + 2\left(\frac{d}{b}\right)\sqrt{1+m^2}\right)^{2/3}}\right)\left(\frac{d}{b}\right)^{5/3} \qquad 19.32(b)$$

12. MOST EFFICIENT CROSS SECTION

- *rectangle*

$$d = \frac{w}{2} \quad \text{[most efficient rectangle]} \qquad 19.34$$

$$A = dw = \frac{w^2}{2} = 2d^2 \qquad 19.35$$

$$P = d + w + d = 2w = 4d \qquad 19.36$$

$$R = \frac{w}{4} = \frac{d}{2} \qquad 19.37$$

- *trapezoid*

$$d = 2R \quad \text{[most efficient trapezoid]} \qquad 19.38$$

$$b = \frac{2d}{\sqrt{3}} \qquad 19.39$$

$$A = \sqrt{3}d^2 \qquad 19.40$$

$$P = 3b = 2\sqrt{3}d \quad \text{[most efficient]} \qquad 19.41$$

$$R = \frac{d}{2} \qquad 19.42$$

14. FLOW MEASUREMENT WITH WEIRS

$$Q = \frac{2}{3}b\sqrt{2g}\left(\left(H + \frac{v_1^2}{2g}\right)^{3/2} - \left(\frac{v_1^2}{2g}\right)^{3/2}\right)$$
$$\begin{bmatrix}\text{rectangular}\\\text{weirs}\end{bmatrix} \qquad 19.47$$

$$Q = \frac{2}{3}b\sqrt{2g}H^{3/2} \quad \begin{bmatrix}\text{rectangular}\\\text{weirs}\end{bmatrix} \qquad 19.48$$

$$Q = \frac{2}{3}C_1b\sqrt{2g}H^{3/2} \quad \begin{bmatrix}\text{rectangular}\\\text{weirs}\end{bmatrix} \qquad 19.49$$

$$C_1 = \left(0.6035 + 0.0813\left(\frac{H}{Y}\right) + \frac{0.000295}{Y}\right)$$
$$\times \left(1 + \frac{0.00361}{H}\right)^{3/2}$$
$$\approx 0.602 + 0.083\left(\frac{H}{Y}\right) \qquad 19.50$$

$$Q \approx 3.33bh^{3/2} \qquad 19.51(b)$$

$$b_{\text{effective}} = b_{\text{actual}} - 0.1NH \qquad 19.52$$

17. BROAD-CRESTED WEIRS AND SPILLWAYS

$$Q = \frac{2}{3}C_1b\sqrt{2g}H^{3/2} \qquad 19.58$$

$$Q = C_{\text{Horton}}b\left(H + \frac{v^2}{2g}\right)^{3/2} \qquad 19.59$$

$$Q = C_sbH^{3/2} \qquad 19.60$$

19. FLOW MEASUREMENT WITH PARSHALL FLUMES

$$Q = KbH_a^n \qquad 19.63$$

$$n = 1.522b^{0.026} \qquad 19.64$$

21. SPECIFIC ENERGY

$$E = d + \frac{v^2}{2g} \qquad 19.66$$

$$E = d + \frac{Q^2}{2gA^2} \quad \text{[general]} \qquad 19.67$$

$$E = d + \frac{Q^2}{2g(wd)^2} \quad \text{[rectangular]} \qquad 19.69$$

25. CRITICAL FLOW AND CRITICAL DEPTH IN RECTANGULAR CHANNELS

$$d_c^3 = \frac{Q^2}{gw^2} \quad \text{[rectangular]} \qquad 19.74$$

$$d_c = \tfrac{2}{3} E_c \qquad 19.75$$

$$v_c = \sqrt{gd_c} \qquad 19.76$$

26. CRITICAL FLOW AND CRITICAL DEPTH IN NONRECTANGULAR CHANNELS

$$\frac{Q^2}{g} = \frac{A^3}{T} \quad \text{[nonrectangular]} \qquad 19.77$$

27. FROUDE NUMBER

$$\text{Fr} = \frac{v}{\sqrt{gL}} \qquad 19.78$$

$$\text{Fr} = \frac{\dfrac{Q}{b}}{\sqrt{gd^3}} \quad \text{[rectangular]} \qquad 19.79$$

$$\text{Fr} = \frac{\dfrac{Q}{b_{\text{ave}}}}{\sqrt{g\left(\dfrac{A}{b_{\text{ave}}}\right)^3}} \quad \text{[nonrectangular]} \qquad 19.80$$

33. HYDRAULIC JUMP

$$d_1 = -\tfrac{1}{2} d_2 + \sqrt{\frac{2v_2^2 d_2}{g} + \frac{d_2^2}{4}} \quad \begin{bmatrix} \text{rectangular} \\ \text{channels} \end{bmatrix} \qquad 19.90$$

$$d_2 = -\tfrac{1}{2} d_1 + \sqrt{\frac{2v_1^2 d_1}{g} + \frac{d_1^2}{4}} \quad \begin{bmatrix} \text{rectangular} \\ \text{channels} \end{bmatrix} \qquad 19.91$$

$$\frac{d_2}{d_1} = \tfrac{1}{2}\left(\sqrt{1 + 8(\text{Fr}_1)^2} - 1\right) \quad \begin{bmatrix} \text{rectangular} \\ \text{channels} \end{bmatrix} \qquad 19.92(a)$$

$$\frac{d_1}{d_2} = \tfrac{1}{2}\left(\sqrt{1 + 8(\text{Fr}_2)^2} - 1\right) \quad \begin{bmatrix} \text{rectangular} \\ \text{channels} \end{bmatrix} \qquad 19.92(b)$$

$$v_1^2 = \left(\frac{gd_2}{2d_1}\right)(d_1 + d_2) \quad \begin{bmatrix} \text{rectangular} \\ \text{channels} \end{bmatrix} \qquad 19.93$$

$$\Delta E = \left(d_1 + \frac{v_1^2}{2g}\right) - \left(d_2 + \frac{v_2^2}{2g}\right) \approx \frac{(d_2 - d_1)^3}{4d_1 d_2} \qquad 19.94$$

38. DETERMINING TYPE OF CULVERT FLOW

A. Type-1 Flow

$$Q = C_d A_c \sqrt{2g\left(h_1 - z + \frac{\alpha v_1^2}{2g} - d_c - h_{f,1\text{-}2}\right)} \qquad 19.101$$

Figure 19.25 *Culvert Flow Classifications*

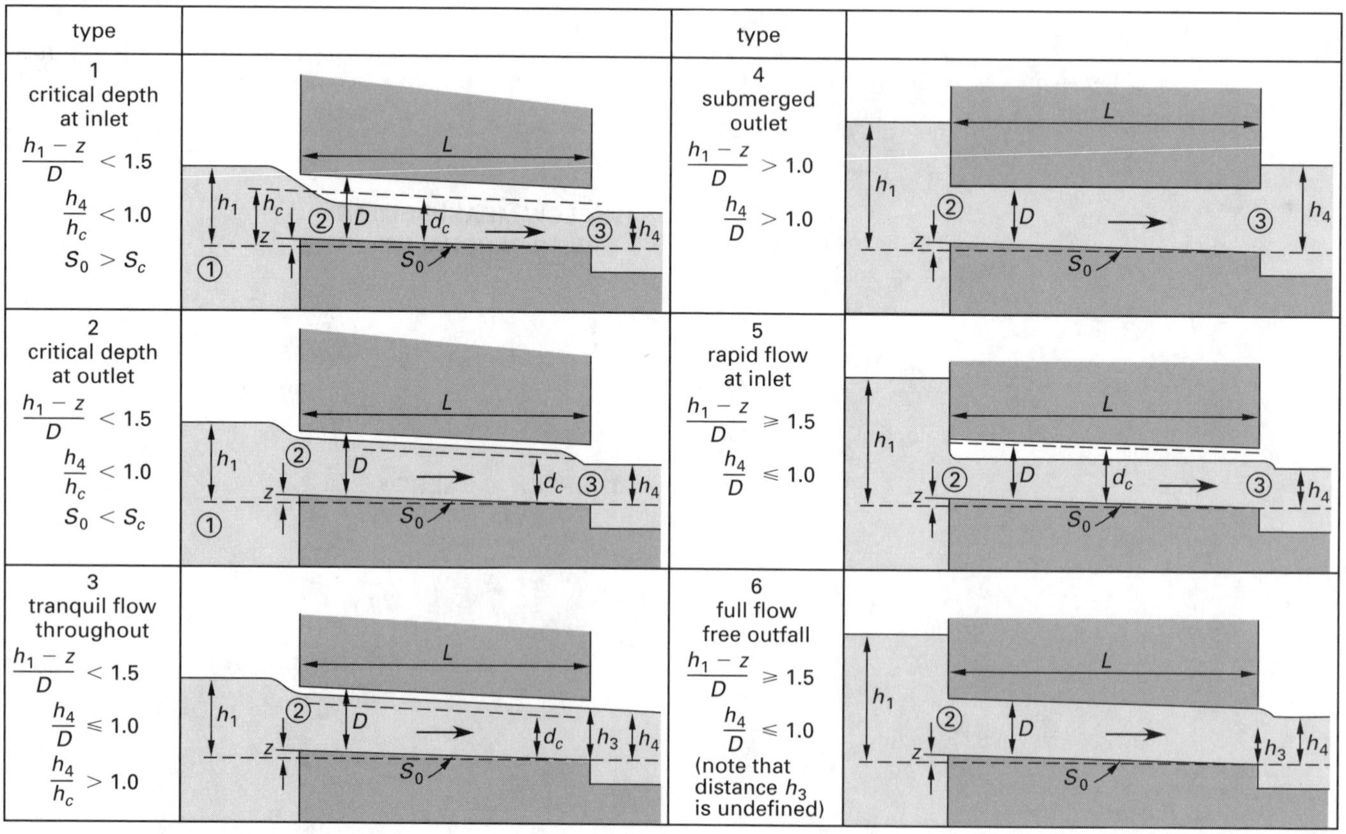

B. Type-2 Flow

$$Q = C_d A_c \sqrt{2g \left(h_1 + \frac{\alpha v_1^2}{2g} - d_c - h_{f,1\text{-}2} - h_{f,2\text{-}3} \right)}$$
19.102

C. Type-3 Flow

$$Q = C_d A_3 \sqrt{2g \left(h_1 + \frac{\alpha v_1^2}{2g} - h_3 - h_{f,1\text{-}2} - h_{f,2\text{-}3} \right)}$$
19.103

D. Type-4 Flow

$$Q = C_d A_o \sqrt{2g \left(\frac{h_1 - h_4}{1 + \dfrac{29 C_d^2 n^2 L}{R^{4/3}}} \right)}$$
19.104

$$Q = C_d A_o \sqrt{2g(h_1 - h_4)}$$
19.105

E. Type-5 Flow

$$Q = C_d A_o \sqrt{2g(h_1 - z)}$$
19.106

F. Type-6 Flow

$$Q = C_d A_o \sqrt{2g(h_1 - h_3 - h_{f,2\text{-}3})}$$
19.107

CERM Chapter 20
Meteorology, Climatology, and Hydrology

Chapter, section, equation, figure, and table numbers correspond to CERM. For additional study material, go to the corresponding chapter and section number in CERM.

5. TIME OF CONCENTRATION

$$t_c = t_{\text{sheet}} + t_{\text{shallow}} + t_{\text{channel}}$$
20.5

6. RAINFALL INTENSITY

$$I = \frac{K}{t_c + b}$$
20.14

7. FLOODS

$$p\{F \text{ event in } n \text{ years}\} = 1 - \left(1 - \frac{1}{F} \right)^n$$
20.20

10. UNIT HYDROGRAPH

$$V = A_d P_{\text{ave,excess}}$$
20.21

15. PEAK RUNOFF FROM THE RATIONAL METHOD

$$Q_p = CIA_d$$
20.36

CERM Chapter 21
Groundwater

Chapter, section, equation, figure, and table numbers correspond to CERM. For additional study material, go to the corresponding chapter and section number in CERM.

2. AQUIFER CHARACTERISTICS

$$w = \frac{m_w}{m_s} = \frac{m_t - m_s}{m_s}$$
21.2

$$n = \frac{V_v}{V_t} = \frac{V_t - V_s}{V_t}$$
21.3

$$e = \frac{V_v}{V_s} = \frac{V_t - V_s}{V_s}$$
21.4

$$e = \frac{n}{1 - n}$$
21.5

$$i = \frac{\Delta H}{L}$$
21.6

3. PERMEABILITY

$$K = \frac{k\gamma}{\mu}$$
21.7

$$K = CD_{\text{mean}}^2$$
21.8

$$K_{\text{cm/s}} \approx C(D_{10,\text{mm}})^2 \quad [0.1 \text{ mm} \leq D_{10,\text{mm}} \leq 3.0 \text{ mm}]$$
21.9

(fine sand $40 \leq C \leq 150$ coarse sand)

4. DARCY'S LAW

$$Q = -KiA_{\text{gross}} = -v_e A_{\text{gross}}$$
21.10

$$q = \frac{Q}{A_{\text{gross}}} = Ki = v_e$$
21.11

$$\text{Re} = \frac{\rho q D_{\text{mean}}}{\mu} = \frac{q D_{\text{mean}}}{\nu}$$
21.12

5. TRANSMISSIVITY

$$T = KY$$
21.13

$$Q = bTi$$
21.14

6. SPECIFIC YIELD, RETENTION, AND CAPACITY

$$S_y = \frac{V_{\text{yielded}}}{V_{\text{total}}} \qquad 21.15$$

$$S_r = \frac{V_{\text{retained}}}{V_{\text{total}}} = n - S_y \qquad 21.16$$

$$\text{specific capacity} = \frac{Q}{s} \qquad 21.17$$

7. DISCHARGE VELOCITY AND SEEPAGE VELOCITY

$$v_{\text{pore}} = \frac{Q}{A_{\text{net}}} = \frac{Q}{nA_{\text{gross}}} = \frac{Q}{nbY} = \frac{Ki}{n} \qquad 21.18$$

$$v_e = nv_{\text{pore}} = \frac{Q}{A_{\text{gross}}} = \frac{Q}{bY} = Ki \qquad 21.19$$

11. WELL DRAWDOWN IN AQUIFERS

$$Q = \frac{\pi K(y_1^2 - y_2^2)}{\ln\left(\frac{r_1}{r_2}\right)} \quad \begin{bmatrix} \text{Dupuit} \\ \text{equation} \end{bmatrix} \qquad 21.25$$

$$Q = \frac{2\pi T(s_2 - s_1)}{\ln\left(\frac{r_1}{r_2}\right)} \quad [s \ll Y] \qquad 21.26$$

$$Q = \frac{2\pi KY(y_1 - y_2)}{\ln\left(\frac{r_1}{r_2}\right)} \quad \begin{bmatrix} \text{Thiem equation;} \\ \text{artesian well} \end{bmatrix} \qquad 21.27$$

12. UNSTEADY FLOW

$$s_{r,t} = \left(\frac{Q}{4\pi KY}\right) W(u)$$

$$= \left(\frac{Q}{4\pi T}\right) W(u) \quad [\text{Theis equation}] \qquad 21.28$$

$$s_1 - s_2 = y_2 - y_1$$

$$= \left(\frac{Q}{4\pi KY}\right)\left(W(u_1) - W(u_2)\right) \qquad 21.31$$

15. SEEPAGE FROM FLOW NETS

$$Q = KH\left(\frac{N_f}{N_p}\right) \quad \begin{bmatrix} \text{per unit} \\ \text{width} \end{bmatrix} \qquad 21.33$$

$$H = H_1 - H_2 \qquad 21.34$$

16. HYDROSTATIC PRESSURE ALONG FLOW PATH

$$p_u = \left(\frac{j}{N_p}\right) H\gamma_w \qquad 21.35$$

$$p_u = \left(\left(\frac{j}{N_p}\right) H + z\right)\gamma_w \qquad 21.36$$

$$U = Np_u A \qquad 21.37$$

$$(FS)_{\text{heave}} = \frac{\text{downward pressure}}{\text{uplift pressure}} \qquad 21.38$$

17. INFILTRATION

$$f_t = f_c + (f_0 - f_c)e^{-kt} \qquad 21.39$$

$$F_t = f_c t + \left(\frac{f_0 - f_c}{k}\right)\left(1 - e^{-kt}\right) \qquad 21.42$$

CERM Chapter 22
Inorganic Chemistry

> Chapter, section, equation, figure, and table numbers correspond to CERM. For additional study material, go to the corresponding chapter and section number in CERM.

9. EQUIVALENT WEIGHT

$$EW = \frac{MW}{\Delta \text{ oxidation number}} \qquad 22.2$$

10. GRAVIMETRIC FRACTION

$$x_i = \frac{m_i}{m_1 + m_2 + \cdots + m_i + \cdots + m_n} = \frac{m_i}{m_t} \qquad 22.3$$

18. UNITS OF CONCENTRATION

F— *formality:* The number of gram formula weights (i.e., molecular weights in grams) per liter of solution.

m— *molality:* The number of gram-moles of solute per 1000 grams of solvent. A "molal" solution contains 1 gram-mole per 1000 grams of solvent.

M— *molarity:* The number of gram-moles of solute per liter of solution. A "molar" (i.e., 1 M) solution contains 1 gram-mole per liter of solution. Molarity is related to normality: $N = M \times \Delta$ oxidation number.

N— *normality:* The number of gram equivalent weights of solute per liter of solution. A solution is "normal" (i.e., 1 N) if there is exactly one gram equivalent weight per liter of solution.

x— *mole fraction:* The number of moles of solute divided by the number of moles of solvent and all solutes.

meq/L— *milligram equivalent weights* of solute *per liter* of solution: calculated by multiplying normality by 1000 or dividing concentration in mg/L by equivalent weight.

mg/L— *milligrams per liter:* The number of milligrams of solute per liter of solution. Same as ppm for solutions of water.

ppm— *parts per million:* The number of pounds (or grams) of solute per million pounds (or grams) of solution. Same as mg/L for solutions of water.

19. pH AND pOH

$$pH = -\log_{10}[H^+] = \log_{10}\left(\frac{1}{[H^+]}\right) \qquad 22.8$$

$$pOH = -\log_{10}[OH^-] = \log_{10}\left(\frac{1}{[OH^-]}\right) \qquad 22.9$$

$$[X] = (\text{fraction ionized}) \times M \qquad 22.10$$

$$pH + pOH = 14 \qquad 22.11$$

26. REVERSIBLE REACTION KINETICS

$$a\text{A} + b\text{B} \longleftrightarrow c\text{C} + d\text{D} \qquad 22.15$$

$$v_{\text{forward}} = k_{\text{forward}}[\text{A}]^a[\text{B}]^b \qquad 22.16$$

$$v_{\text{reverse}} = k_{\text{reverse}}[\text{C}]^c[\text{D}]^d \qquad 22.17$$

27. EQUILIBRIUM CONSTANT

$$K = \frac{[\text{C}]^c[\text{D}]^d}{[\text{A}]^a[\text{B}]^b} = \frac{k_{\text{forward}}}{k_{\text{reverse}}} \qquad 22.19$$

28. IONIZATION CONSTANT

$$K_{\text{ionization}} = \frac{MX^2}{1-X} \qquad 22.24$$

32. ENTHALPY OF REACTION

$$\Delta H_r = \sum \Delta H_{f,\text{products}} - \sum \Delta H_{f,\text{reactants}} \qquad 22.26$$

Table of Relative Atomic Weights
(based on the atomic mass of $^{12}C = 12$)

name	symbol	atomic number	atomic weight	name	symbol	atomic number	atomic weight
actinium	Ac	89	–	mercury	Hg	80	200.59
aluminum	Al	13	26.9815	molybdenum	Mo	42	95.94
americium	Am	95	–	neodymium	Nd	60	144.24
antimony	Sb	51	121.75	neon	Ne	10	20.183
argon	Ar	18	39.948	neptunium	Np	93	–
arsenic	As	33	74.9216	nickel	Ni	28	58.71
astatine	At	85	–	niobium	Nb	41	92.906
barium	Ba	56	137.34	nitrogen	N	7	14.0067
berkelium	Bk	97	–	nobelium	No	102	
beryllium	Be	4	9.0122	osmium	Os	76	190.2
bismuth	Bi	83	208.980	oxygen	O	8	15.9994
boron	B	5	10.811	palladium	Pd	46	106.4
bromine	Br	35	79.904	phosphorus	P	15	30.9738
cadmium	Cd	48	112.40	platinum	Pt	78	195.09
calcium	Ca	20	40.08	plutonium	Pu	94	–
californium	Cf	98	–	polonium	Po	84	–
carbon	C	6	12.01115	potassium	K	19	39.102
cerium	Ce	58	140.12	praseodymium	Pr	59	140.907
cesium	Cs	55	132.905	promethium	Pm	61	–
chlorine	Cl	17	35.453	protactinium	Pa	91	–
chromium	Cr	24	51.996	radium	Ra	88	–
cobalt	Co	27	58.9332	radon	Rn	86	–
copper	Cu	29	63.546	rhenium	Re	75	186.2
curium	Cm	96	–	rhodium	Rh	45	102.905
dysprosium	Dy	66	162.50	rubidium	Rb	37	85.47
einsteinium	Es	99	–	ruthenium	Ru	44	101.07
erbium	Er	68	167.26	samarium	Sm	62	150.35
europium	Eu	63	151.96	scandium	Sc	21	44.956
fermium	Fm	100	–	selenium	Se	34	78.96
fluorine	F	9	18.9984	silicon	Si	14	28.086
francium	Fr	87	–	silver	Ag	47	107.868
gadolinium	Gd	64	157.25	sodium	Na	11	22.9898
gallium	Ga	31	69.72	strontium	Sr	38	87.62
germanium	Ge	32	72.59	sulfur	S	16	32.064
gold	Au	79	196.967	tantalum	Ta	73	180.948
hafnium	Hf	72	178.49	technetium	Tc	43	–
helium	He	2	4.0026	tellurium	Te	52	127.60
holmium	Ho	67	164.930	terbium	Tb	65	158.924
hydrogen	H	1	1.00797	thallium	Tl	81	204.37
indium	In	49	114.82	thorium	Th	90	232.038
iodine	I	53	126.9044	thulium	Tm	69	168.934
iridium	Ir	77	192.2	tin	Sn	50	118.69
iron	Fe	26	55.847	titanium	Ti	22	47.90
krypton	Kr	36	83.80	tungsten	W	74	183.85
lanthanum	La	57	138.91	uranium	U	92	238.03
lead	Pb	82	207.19	vanadium	V	23	50.942
lithium	Li	3	6.939	xenon	Xe	54	131.30
lutetium	Lu	71	174.97	ytterbium	Yb	70	173.04
magnesium	Mg	12	24.312	yttrium	Y	39	88.905
manganese	Mn	25	54.9380	zinc	Zn	30	65.37
mendelevium	Md	101	–	zirconium	Zr	40	91.22

CERM Chapter 24
Combustion and Incineration

Chapter, section, equation, figure, and table numbers correspond to CERM. For additional study material, go to the corresponding chapter and section number in CERM.

6. MOISTURE

$$G_{H,combined} = \frac{G_O}{8} \qquad 24.1$$

$$G_{H,available} = G_{H,total} - \frac{G_O}{8} \qquad 24.2$$

24. ATMOSPHERIC AIR

Table 24.6 Composition of Dry Air [a]

component	percent by weight	percent by volume
oxygen	23.15	20.95
nitrogen/inerts	76.85	79.05
ratio of nitrogen to oxygen	3.320	3.773[b]
ratio of air to oxygen	4.320	4.773

[a]Inert gases and CO_2 included as N_2.
[b]The value is also reported by various sources as 3.76, 3.78, and 3.784.

25. COMBUSTION REACTIONS

Table 24.7 Ideal Combustion Reactions

fuel	formula	reaction equation (excluding nitrogen)
carbon (to CO)	C	$2C + O_2 \longrightarrow 2CO$
carbon (to CO_2)	C	$C + O_2 \longrightarrow CO_2$
sulfur (to SO_2)	S	$S + O_2 \longrightarrow SO_2$
sulfur (to SO_3)	S	$2S + 3O_2 \longrightarrow 2SO_3$
carbon monoxide	CO	$2CO + O_2 \longrightarrow 2CO_2$
methane	CH_4	$CH_4 + 2O_2 \longrightarrow CO_2 + 2H_2O$
acetylene	C_2H_2	$2C_2H_2 + 5O_2 \longrightarrow 4CO_2 + 2H_2O$
ethylene	C_2H_4	$C_2H_4 + 3O_2 \longrightarrow 2CO_2 + 2H_2O$
ethane	C_2H_6	$2C_2H_6 + 7O_2 \longrightarrow 4CO_2 + 6H_2O$
hydrogen	H_2	$2H_2 + O_2 \longrightarrow 2H_2O$
hydrogen sulfide	H_2S	$2H_2S + 3O_2 \longrightarrow 2H_2O + 2SO_2$
propane	C_3H_8	$C_3H_8 + 5O_2 \longrightarrow 3CO_2 + 4H_2O$
n-butane	C_4H_{10}	$2C_4H_{10} + 13O_2 \longrightarrow 8CO_2 + 10H_2O$
octane	C_8H_{18}	$2C_8H_{18} + 25O_2 \longrightarrow 16CO_2 + 18H_2O$
olefin series	C_nH_{2n}	$2C_nH_{2n} + 3nO_2 \longrightarrow 2nCO_2 + 2nH_2O$
paraffin series	C_nH_{2n+2}	$2C_nH_{2n+2} + (3n+1)O_2 \longrightarrow 2nCO_2 + (2n+2)H_2O$

(Multiply oxygen volumes by 3.773 to get nitrogen volumes.)

27. STOICHIOMETRIC AIR

$$R_{a/f,ideal} = \frac{m_{air,ideal}}{m_{fuel}} \qquad 24.5$$

$$R_{a/f,ideal} = (34.5)\left(\frac{G_C}{3} + G_H - \frac{G_O}{8} + \frac{G_S}{8}\right) \qquad \text{[solid fuels]} \qquad 24.6$$

$$R_{a/f,ideal} = \sum J_i G_i \quad \text{[gaseous fuels]} \qquad 24.7$$

$$\text{volumetric air/fuel ratio} = \sum K_i B_i \quad \text{[gaseous fuels]} \qquad 24.8$$

30. ACTUAL AND EXCESS AIR

$$R_{a/f,\text{actual}} = \frac{m_{\text{air,actual}}}{m_{\text{fuel}}}$$

$$= \frac{3.04 B_{\text{N}_2} \left(G_{\text{C}} + \dfrac{G_{\text{S}}}{1.833} \right)}{B_{\text{CO}_2} + B_{\text{CO}}} \qquad 24.9$$

34. DEW POINT OF FLUE GAS MOISTURE

$$\text{partial pressure} = (\text{water vapor mole fraction})$$
$$\times (\text{flue gas pressure}) \qquad 24.13$$

35. HEAT OF COMBUSTION

$$\text{HHV} = \text{LHV} + m_{\text{water}} h_{fg} \qquad 24.14$$

$$G_{\text{H,available}} = G_{\text{H,total}} - \frac{G_{\text{O}}}{8} \qquad 24.15$$

$$\text{HHV}_{\text{Btu/lbm}} = 14{,}093 G_{\text{C}} + 60{,}958 \left(G_{\text{H}} - \frac{G_{\text{O}}}{8} \right)$$
$$+ 3983 G_{\text{S}} \text{ solid fuels}$$
$$\qquad 24.16(b)$$

$$\text{HHV}_{\text{gasoline,Btu/lbm}} = 18{,}320 + 40(^\circ\text{Baumé} - 10)$$
$$\qquad 24.17(b)$$

$$\text{HHV}_{\text{fuel oil,Btu/lbm}} = 22{,}320 - 3780(\text{SG})^2$$
$$\qquad 24.18(b)$$

36. MAXIMUM THEORETICAL COMBUSTION (FLAME) TEMPERATURE

$$T_{\text{max}} = T_i + \frac{\text{lower heat of combustion}}{m_{\text{products}} c_{p,\text{mean}}} \qquad 24.19$$

37. COMBUSTION LOSSES

$$q_1 = m_{\text{flue gas}} c_p (T_{\text{flue gas}} - T_{\text{incoming air}}) \qquad 24.20$$

$$q_2 = m_{\text{vapor}}(h_g - h_f) = 8.94 G_{\text{H}}(h_g - h_f) \qquad 24.21$$

$$q_3 = m_{\text{atmospheric water vapor}}(h_g - h'_g)$$
$$= \omega m_{\text{combustion air}}(h_g - h'_g) \qquad 24.22$$

$$q_4 = \frac{(\text{HHV}_{\text{C}} - \text{HHV}_{\text{CO}})G_{\text{C}} B_{\text{CO}}}{B_{\text{CO}_2} + B_{\text{CO}}} \qquad 24.23$$

$$q_5 = \text{HHV}_{\text{C}} m_{\text{ash}} G_{\text{C,ash}} \qquad 24.24$$

38. COMBUSTION EFFICIENCY

$$\eta = \frac{\text{useful heat extracted}}{\text{heating value}}$$

$$= \frac{m_{\text{steam}}(h_{\text{steam}} - h_{\text{feedwater}})}{\text{HHV}} \qquad 24.25$$

$$\eta = \frac{\text{HHV} - q_1 - q_2 - q_3 - q_4 - q_5 - \text{radiation}}{\text{HHV}}$$

$$= \frac{\text{LHV} - q_1 - q_4 - q_5 - \text{radiation}}{\text{HHV}} \qquad 24.26$$

CERM Chapter 25
Water Supply Quality and Testing

Chapter, section, equation, figure, and table numbers correspond to CERM. For additional study material, go to the corresponding chapter and section number in CERM.

6. HARDNESS AND ALKALINITY

Figure 25.1 Hardness and Alkalinity[a,b]

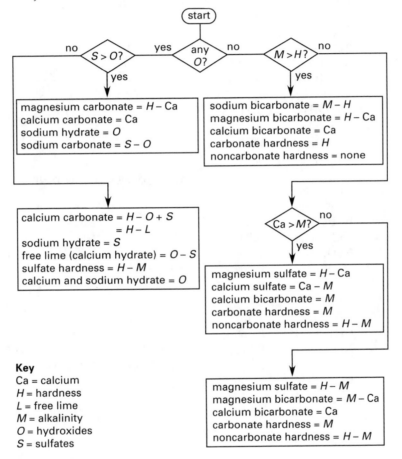

[a] All concentrations are expressed as $CaCO_3$.

[b] Not for use when other ionic species are present in significant quantities.

Conversions From mg/L as a Substance to mg/L as CaCO₃

Multiply the mg/L of the substances listed in the following table by the corresponding factors to obtain mg/L as $CaCO_3$. For example, 70 mg/L of Mg^{++} would be $(70 \text{ mg/L})(4.10) = 287$ mg/L as $CaCO_3$.

substance	factor	substance	factor
Al^{+++}	5.56	HCO_3^-	0.82
$Al_2(SO_4)_3$	0.88[a]	K^+	1.28
$AlCl_3$	1.13	KCl	0.67
$Al(OH)_3$	1.92	K_2CO_3	0.72
Ba^{++}	0.73	Mg^{++}	4.10
$Ba(OH)_2$	0.59	$MgCl_2$	1.05
$BaSO_4$	0.43	$MgCO_3$	1.19
Ca^{++}	2.50	$Mg(HCO_3)_2$	0.68
$CaCl_2$	0.90	MgO	2.48
$CaCO_3$	1.00	$Mg(OH)_2$	1.71
$Ca(HCO_3)_2$	0.62	$Mg(NO_3)_2$	0.67
CaO	1.79	$MgSO_4$	0.83
$Ca(OH)_2$	1.35	Mn^{++}	1.82
$CaSO_4$	0.74[a]	Na^+	2.18
Cl^-	1.41	$NaCl$	0.85
CO_2	2.27	Na_2CO_3	0.94
CO_3^{--}	1.67	$NaHCO_3$	0.60
Cu^{++}	1.57	$NaNO_3$	0.59
Cu^{+++}	2.36	$NaOH$	1.25
$CuSO_4$	0.63	Na_2SO_4	0.70[a]
F^-	2.66	NH_3	2.94
Fe^{++}	1.79	NH_4^+	2.78
Fe^{+++}	2.69	NH_4OH	1.43
$Fe(OH)_3$	1.41	$(NH_4)_2SO_4$	0.76
$FeSO_4$	0.66[a]	NO_3^-	0.81
$Fe_2(SO_4)_3$	0.75	OH^-	2.94
$FeCl_3$	0.93	PO_4^{---}	1.58
H^+	50.0	SO_4^{--}	1.04
		Zn^{++}	1.54

[a] anhydrous

CERM Chapter 26
Water Supply Treatment and Distribution

> Chapter, section, equation, figure, and table numbers correspond to CERM. For additional study material, go to the corresponding chapter and section number in CERM.

11. SEDIMENTATION PHYSICS

$$\text{Re} = \frac{v_s D}{\nu} \qquad 26.2$$

$$v_{s,\text{ft/sec}} = \frac{(\rho_{\text{particle}} - \rho_{\text{water}})D_{\text{ft}}^2 g}{18\mu g_c} \left[\begin{array}{c} \text{settling} \\ \text{velocity} \end{array} \right]$$

$$= \frac{(\text{SG}_{\text{particle}} - 1)D_{\text{ft}}^2 g}{18\nu} \quad \text{[Stokes' law]} \quad 26.3(b)$$

$$v_s = \sqrt{\frac{4gD(\text{SG}_{\text{particle}} - 1)}{3C_D}} \qquad 26.4$$

12. SEDIMENTATION TANKS

$$t_{\text{settling}} = \frac{h}{v_s} \qquad 26.5$$

$$t_d = \frac{V_{\text{tank}}}{Q} = \frac{Ah}{Q} \qquad 26.6$$

$$v^* = \frac{Q_{\text{filter}}}{A_{\text{surface}}} = \frac{Q_{\text{filter}}}{bL} \qquad 26.7$$

15. DOSES OF COAGULANTS AND OTHER COMPOUNDS

$$F_{\text{lbm/day}} = \frac{D_{\text{mg/L}} Q_{\text{MGD}} \left(8.345 \dfrac{\text{lbm-L}}{\text{mg·MG}} \right)}{PG}$$

$$26.11(b)$$

16. MIXERS AND MIXING KINETICS

$$V = tQ \qquad 26.12$$

$$t_{\text{complete}} = \frac{V}{Q} = \frac{1}{K}\left(\frac{C_i}{C_o} - 1 \right) \qquad 26.13$$

$$t_{\text{plug flow}} = \frac{V}{Q}$$

$$= \frac{L}{v_{\text{flow through}}} = \frac{1}{K}\ln\left(\frac{C_i}{C_o} \right) \qquad 26.14$$

17. MIXING PHYSICS

$$F_D = \frac{C_D A \rho v_{\text{mixing}}^2}{2g_c} = \frac{C_D A \gamma v_{\text{mixing}}^2}{2g}$$

26.15(b)

$$v_{\text{paddle,ft/sec}} = \frac{2\pi R n_{\text{rpm}}}{60 \dfrac{\text{sec}}{\text{min}}}$$

26.16

$$v_{\text{mixing}} = v_{\text{paddle}} - v_{\text{water}}$$

26.17

$$P_{\text{hp}} = \frac{F_D v_{\text{mixing}}}{550 \dfrac{\text{ft-lbf}}{\text{hp-sec}}}$$

$$= \frac{C_D A \gamma v_{\text{mixing}}^3}{2g \left(550 \dfrac{\text{ft-lbf}}{\text{hp-sec}}\right)}$$

26.19(b)

$$G = \sqrt{\frac{P}{\mu V_{\text{tank}}}}$$

26.20

$$P = \mu G^2 V_{\text{tank}}$$

26.21

$$Gt_d = \frac{V_{\text{tank}}}{Q} \sqrt{\frac{P}{\mu V_{\text{tank}}}}$$

$$= \frac{1}{Q} \sqrt{\frac{P V_{\text{tank}}}{\mu}}$$

26.22

18. IMPELLER CHARACTERISTICS

$$\text{Re} = \frac{D^2 n \rho}{g_c \mu}$$

26.23(b)

$$P = \frac{N_P n^3 D^5 \rho}{g_c} \quad \begin{bmatrix} \text{geometrically} \\ \text{similar; turbulent} \end{bmatrix}$$

26.24(b)

$$P = \frac{\rho g Q h_v}{g_c} \quad \begin{bmatrix} \text{geometrically} \\ \text{similar; turbulent} \end{bmatrix}$$

26.25(b)

$$Q = N_Q n D^3 \quad \begin{bmatrix} \text{geometrically} \\ \text{similar; turbulent} \end{bmatrix}$$

26.26

$$h_v = \frac{N_P n^2 D^2}{N_Q g} \quad \begin{bmatrix} \text{geometrically} \\ \text{similar; turbulent} \end{bmatrix}$$

26.27

21. SLUDGE QUANTITIES

$$V_{\text{sludge}} = \frac{m_{\text{sludge}}}{\rho_{\text{water}} (\text{SG})_{\text{sludge}} G}$$

26.28

$$m_{\text{sludge,lbm/day}} =$$
$$\left(8.345 \frac{\text{lbm-L}}{\text{mg-MG}}\right)$$
$$\times (Q_{\text{MGD}})(0.46 D_{\text{alum,mg/L}} + \Delta \text{TSS}_{\text{mg/L}} + M_{\text{mg/L}})$$

26.29(b)

$$V_{2,\text{sludge}} \approx V_{1,\text{sludge}} \left(\frac{G_1}{G_2}\right)$$

26.30

22. FILTRATION

$$\text{loading rate} = \frac{Q}{A}$$

26.31

23. FILTER BACKWASHING

$$V = A_{\text{filter}}(\text{rate of rise})t_{\text{backwash}}$$

26.32

26. FLUORIDATION

$$F_{\text{lbm/day}} = \frac{C_{\text{mg/L}} Q_{\text{MGD}} \left(8.345 \dfrac{\text{lbm-L}}{\text{mg-MG}}\right)}{PG}$$

26.33(b)

28. TASTE AND ODOR CONTROL

$$\text{TON} = \frac{V_{\text{raw sample}} + V_{\text{dilution water}}}{V_{\text{raw sample}}}$$

31. WATER SOFTENING BY ION EXCHANGE

$$V_{\text{water}} = \frac{(\text{specific working capacity})(V_{\text{exchange material}})}{\text{hardness}}$$

26.42

32. REGENERATION OF ION EXCHANGE RESINS

$$m_{\text{salt}} = (\text{specific working capacity})(V_{\text{exchange material}})$$
$$\times (\text{salt requirement})$$

26.43

33. STABILIZATION

$$\text{LI} = \text{pH} - \text{pH}_{\text{sat}}$$

26.44

$$\text{pH}_{\text{sat}} = (\text{pK}_2 - \text{pK}_1) + \text{pCa} + \text{pM}$$

$$= -\log\left(\frac{K_2}{K_1}\right)[\text{Ca}^{++}][M]$$

26.45

$$\text{RI} = 2\text{pH}_{\text{sat}} - \text{pH}$$

26.46

40. WATER DEMAND

$$Q_{\text{instantaneous}} = M(\text{AADF})$$

26.51

41. FIRE FIGHTING DEMAND

$$Q_{\text{gpm}} = 18F\sqrt{A_{\text{ft}^2}}$$

26.52

$$Q_{\text{gpm}} = 1020\sqrt{P}\left(1 - 0.01\sqrt{P}\right)$$

26.53

Environmental

CERM Chapter 28
Wastewater Quantity and Quality

Chapter, section, equation, figure, and table numbers correspond to CERM. For additional study material, go to the corresponding chapter and section number in CERM.

4. WASTEWATER QUANTITY

$$\frac{Q_{\text{peak}}}{Q_{\text{ave}}} = \frac{18 + \sqrt{P}}{4 + \sqrt{P}} \quad \begin{bmatrix} \text{TSS equation;} \\ P \text{ in thousands} \end{bmatrix} \qquad 28.1$$

15. MICROBIAL GROWTH

$$r_g = \frac{dX}{dt} = \mu X \qquad 28.7$$

$$\mu = \mu_m \left(\frac{S}{K_s + S} \right) \qquad 28.8$$

$$r_g = \frac{\mu_m X S}{K_s + S} \qquad 28.9$$

$$r_{su} = \frac{dS}{dt} = \frac{-\mu_m X S}{Y(K_s + S)} \qquad 28.10$$

$$k = \frac{\mu_m}{Y} \qquad 28.11$$

$$\frac{dS}{dt} = \frac{-kXS}{K_s + S} \qquad 28.12$$

$$r_d = -k_d X \qquad 28.13$$

$$r_g' = \frac{\mu_m X S}{K_s + S} - k_d X$$
$$= -Y r_{su} - k_d X \qquad 28.14$$

$$\mu' = \mu_m \left(\frac{S}{K_s + S} \right) - k_d \qquad 28.15$$

16. DISSOLVED OXYGEN IN WASTEWATER

$$D = \text{DO}_{\text{sat}} - \text{DO} \qquad 28.16$$

17. REOXYGENATION

$$r_r = K_r(\text{DO}_{\text{sat}} - \text{DO}) \qquad 28.17$$

$$D_t = D_0 10^{-K_r t} \quad [\text{base-10}] \qquad 28.18$$

$$D_t = D_0 e^{-K_r' t} \quad [\text{base-}e] \qquad 28.19$$

$$K_r' = 2.303 K_r \qquad 28.20$$

$$K_{r,68°\text{F}}' \approx \frac{12.9\sqrt{\text{v}_{\text{ft/sec}}}}{d_{\text{ft}}^{1.5}} \quad \begin{bmatrix} \text{O'Connor} \\ \text{and Dobbins} \end{bmatrix}$$
$$\begin{bmatrix} 1 \text{ ft} < d < 30 \text{ ft} \\ 0.5 \text{ ft/sec} < \text{v} < 1.6 \text{ ft/sec} \end{bmatrix}$$
$$28.21(b)$$

$$K_{r,68°\text{F}}' = \frac{11.61(\text{v}_{\text{ft/sec}})^{0.969}}{d_{\text{ft}}^{1.67}}$$
$$\begin{bmatrix} 2 \text{ ft} < d < 11 \text{ ft} \\ 1.8 \text{ ft/sec} < \text{v} < 5 \text{ ft/sec} \end{bmatrix}$$
$$28.22(b)$$

$$K_{r,T}' = K_{r,20°\text{C}}'(1.024)^{T-20°\text{C}} \qquad 28.23$$

$$K_{r,T_1}' = K_{r,T_2}' \theta_r^{T_1 - T_2} \qquad 28.24$$

18. DEOXYGENATION

$$r_{d,t} = -K_d \text{DO} \qquad 28.25$$

$$K_{d,T}' = K_{d,20°\text{C}}' \theta^{T-20°\text{C}} \qquad 28.26$$

$$K_{d,T_1}' = K_{d,T_2}' \theta_d^{T_1 - T_2} \qquad 28.27$$

20. BIOCHEMICAL OXYGEN DEMAND

$$\text{BOD}_5 = \frac{\text{DO}_i - \text{DO}_f}{\dfrac{\text{V}_{\text{sample}}}{\text{V}_{\text{sample}} + \text{V}_{\text{dilution}}}} \qquad 28.28$$

$$\text{BOD}_t = \text{BOD}_u(1 - 10^{-K_d t}) \qquad 28.29$$

$$\text{BOD}_u \approx 1.463 \text{BOD}_5 \qquad 28.30$$

$$\text{BOD}_{T°\text{C}} = \text{BOD}_{20°\text{C}}(0.02 T°\text{C} + 0.6) \qquad 28.31$$

21. SEEDED BOD

$$\text{BOD} = \frac{\text{DO}_i - \text{DO}_f - x(\text{DO}_i^* - \text{DO}_f^*)}{\dfrac{V_{\text{sample}}}{V_{\text{sample}} + V_{\text{dilution}}}} \qquad 28.32$$

22. DILUTION PURIFICATION

$$C_f = \frac{C_1 Q_1 + C_2 Q_2}{Q_1 + Q_2} \qquad 28.33$$

23. RESPONSE TO DILUTION PURIFICATION

$$D_t = \left(\frac{K_d \text{BOD}_u}{K_r - K_d}\right)(10^{-K_d t} - 10^{-K_r t}) + D_0(10^{-K_r t}) \qquad 28.34$$

$$x_c = v t_c \qquad 28.35$$

$$t_c = \left(\frac{1}{K_r - K_d}\right)$$
$$\times \log_{10}\left(\left(\frac{K_d \text{BOD}_u - K_r D_0 + K_d D_0}{K_d \text{BOD}_u}\right)\left(\frac{K_r}{K_d}\right)\right) \qquad 28.36$$

$$D_c = \left(\frac{K_d \text{BOD}_u}{K_r}\right) 10^{-K_d t_c} \qquad 28.37$$

Figure 28.5 *Oxygen Sag Curve*

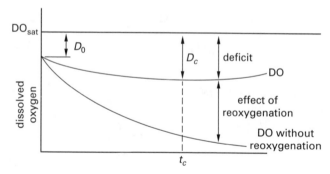

CERM Chapter 29
Wastewater Treatment: Equipment and Processes

Chapter, section, equation, figure, and table numbers correspond to CERM. For additional study material, go to the corresponding chapter and section number in CERM.

11. AERATED LAGOONS

$$t_d = \frac{V}{Q} = k_{1,\text{base}-e}^{1-\eta} \qquad 29.1$$

$$k_{1,\text{base}-e} = 2.3 k_{1,\text{base}-10} \qquad 29.2$$

13. GRIT CHAMBERS

$$v = \sqrt{8k\left(\frac{g d_p}{f}\right)(\text{SG}_p - 1)} \qquad 29.3$$

17. PLAIN SEDIMENTATION BASINS/CLARIFIERS

$$v^* = \frac{Q}{A} \qquad 29.4$$

$$t_d = \frac{V}{Q} \qquad 29.5$$

$$\text{weir loading} = \frac{Q}{L} \qquad 29.6$$

19. TRICKLING FILTERS

$$S_{ps} + RS_o = (1+R)S_i \qquad 29.7$$

$$S_i = \frac{S_{ps} + RS_o}{1+R} \qquad 29.8$$

$$\eta = \frac{S_{\text{removed}}}{S_{ps}} = \frac{S_{ps} - S_o}{S_{ps}} \qquad 29.9$$

$$R = \frac{Q_r}{Q_w} \qquad 29.10$$

$$L_H = \frac{Q_w + Q_r}{A} = \frac{Q_w(1+R)}{A} \qquad 29.11$$

$$L_{\text{BOD}} = \frac{Q_{w,\text{MGD}} S_{\text{mg/L}}\left(8.345\ \dfrac{\text{lbm-L}}{\text{MG-mg}}\right)(1000)}{V_{\text{ft}^3}} \qquad 29.12$$

21. NATIONAL RESEARCH COUNCIL EQUATION

$$\eta = \frac{1}{1 + 0.0561\sqrt{\dfrac{L_{\text{BOD}}}{F}}} \qquad 29.13$$

$$\eta = \frac{1}{1 + 0.0085\sqrt{\dfrac{L_{\text{BOD,lbm/day}}}{V_{\text{ac-ft}} F}}} \qquad 29.14$$

$$F = \frac{1+R}{(1+wR)^2} \qquad 29.15$$

$$\eta_2 = \frac{1}{1 + \left(\dfrac{0.0561}{1-\eta_1}\right)\sqrt{\dfrac{L_{\text{BOD}}}{F}}} \quad \begin{bmatrix}\text{two-stage}\\\text{with clarifier}\end{bmatrix} \qquad 29.16$$

22. VELZ EQUATION

$$\frac{\Delta S_i}{S_i} = \frac{S_i - S_o}{S_i} = 10^{-KZ} \qquad 29.17$$

$$K_T = K_{20°C}(1.047)^{T-20°C} \qquad 29.18$$

CERM Chapter 30
Activated Sludge and Sludge Processing

Chapter, section, equation, figure, and table numbers correspond to CERM. For additional study material, go to the corresponding chapter and section number in CERM.

5. SLUDGE PARAMETERS

$$F = S_o Q_o \qquad 30.1$$

$$M = V_a X \qquad 30.2$$

$$\text{F:M} = \frac{S_{o,\text{mg/L}} Q_{o,\text{MGD}}}{V_{a,\text{MG}} X_{\text{mg/L}}} \qquad 30.3$$

$$= \frac{S_{o,\text{mg/L}}}{\theta_{\text{days}} X_{\text{mg/L}}}$$

$$\text{F:M} = \frac{S_{o,\text{mg/L}} Q_{o,\text{MGD}}}{V_{a,\text{MG}} \text{MLSS}_{\text{mg/L}}} \qquad 30.4$$

$$\theta_c = \frac{V_a X}{Q_e X_e + Q_w X_w} \qquad 30.5$$

$$\theta_{\text{BOD}} = \frac{1}{\text{F:M}}$$

$$= \frac{V_{a,\text{MG}} X_{\text{mg/L}}}{S_{o,\text{mg/L}} Q_{o,\text{MGD}}} \qquad 30.6$$

$$\text{SVI} = \frac{\left(1000 \frac{\text{mg}}{\text{g}}\right) V_{\text{settled,mL/L}}}{\text{MLSS}_{\text{mg/L}}} \qquad 30.7$$

$$\text{TSS}_{\text{mg/L}} = \frac{\left(1000 \frac{\text{mg}}{\text{g}}\right)\left(1000 \frac{\text{mL}}{\text{L}}\right)}{\text{SVI}_{\text{mL/g}}} \qquad 30.8$$

6. SOLUBLE BOD ESCAPING TREATMENT

$$\text{BOD}_e = \text{BOD}_{\text{escaping treatment}}$$
$$+ \text{BOD}_{\text{effluent suspended solids}}$$
$$= S + S_e$$
$$= S + 1.42 f G X_e \qquad 30.9$$

$$f = \frac{\text{BOD}_5}{\text{BOD}_u} \qquad 30.10$$

7. PROCESS EFFICIENCY

$$\eta_{\text{BOD}} = \frac{S_o - S}{S_o} \qquad 30.11$$

8. PLUG FLOW AND STIRRED TANK MODELS

$$\frac{1}{\theta_c} = \frac{\mu_m(S_o - S)}{S_o - S + (1+R)K_s \ln\left(\frac{S_i}{S}\right)} - k_d$$
$$[\text{PFR only}] \qquad 30.12$$

$$\mu_m = kY \qquad 30.13$$

$$S_i = \frac{S_o + RS}{1+R} \quad [\text{PFR only}] \qquad 30.14$$

$$r_{su} = \frac{-\mu_m S X}{Y(K_s + S)} \quad [\text{PFR only}] \qquad 30.15$$

$$U = \eta(\text{F:M}) = \frac{-r_{su}}{X}$$
$$= \frac{S_o - S}{\theta X} \qquad 30.16$$

$$\frac{1}{\theta_c} = Y(\text{F:M})\eta - k_d$$
$$= -Y\left(\frac{r_{su}}{X}\right) - k_d$$
$$= YU - k_d \quad [\text{CSTR only}] \qquad 30.17$$

$$S = \frac{K_s(1 + k_d \theta_c)}{\theta_c(\mu_m - k_d) - 1} \quad [\text{CSTR only}] \qquad 30.18$$

$$V_a = \theta Q_o$$
$$= \frac{\theta_c Q_o Y(S_o - S)}{X(1 + k_d \theta_c)}$$
$$[\text{PFR and CSTR}] \qquad 30.19$$

$$\theta = \frac{V_a}{Q_o} \quad [\text{PFR and CSTR}] \qquad 30.20$$

$$\theta_s = \frac{V_a + V_s}{Q_o} \quad [\text{PFR and CSTR}] \qquad 30.21$$

$$X = \frac{\left(\frac{\theta_c}{\theta}\right) Y(S_o - S)}{1 + k_d \theta_c}$$
$$[\text{PFR and CSTR}] \qquad 30.22$$

$$Y_{\text{obs}} = \frac{Y}{1 + k_d \theta_c} \quad [\text{PFR and CSTR}] \qquad 30.23$$

$$P_{x,\text{kg/day}} = \frac{Q_{w,\text{m}^3/\text{day}} X_{r,\text{mg/L}}}{1000 \frac{\text{g}}{\text{kg}}}$$
$$= \frac{Y_{\text{obs,mg/mg}} Q_{o,\text{m}^3/\text{day}}(S_o - S)_{\text{mg/L}}}{1000 \frac{\text{g}}{\text{kg}}}$$
$$[\text{PFR and CSTR}] \qquad 30.24$$

$$m_{e,\text{solids}} = Q_e X_e \quad \text{[PFR and CSTR]} \qquad 30.25$$

$$\theta_c = \frac{V_a X}{Q_w X_r + Q_e X_e}$$

$$= \frac{V_a X}{Q_w X_r + (Q_o - Q_w) X_e} \quad \text{[PFR and CSTR]}$$

$$30.26$$

$$\theta_c \approx \frac{V_a X}{Q_w X_r} \quad \text{[PFR and CSTR]} \qquad 30.27$$

$$\theta_c \approx \frac{V_a}{Q_w} \quad \text{[CSTR]} \qquad 30.28$$

10. AERATION TANKS

$$\theta = \frac{V_a}{Q_o} \qquad 30.29$$

$$L_{\text{BOD}} = \frac{S_o Q_o}{V_a \left(1000 \; \dfrac{\text{mg·m}^3}{\text{kg·L}}\right)} \qquad 30.30$$

$$L_{\text{BOD}} = \frac{S_{o,\text{mg/L}} Q_{o,\text{MGD}} \times \left(8.345 \; \dfrac{\text{lbm-L}}{\text{MG-mg}}\right)(1000)}{V_{a,\text{ft}^3}} \qquad 30.31$$

$$\dot{m}_{\text{oxygen}} = K_t D \qquad 30.32$$

$$D = \beta \text{DO}_{\text{saturated water}} - \text{DO}_{\text{mixed liquor}} \qquad 30.33$$

$$\dot{m}_{\text{oxygen,kg/day}} = \left(\frac{Q_{o,\text{m}^3/\text{day}} (S_o - S)_{\text{mg/L}}}{f \left(1000 \; \dfrac{\text{mg·m}^3}{\text{kg·L}}\right)}\right) - 1.42 P_{x,\text{kg/day}} \quad \text{[PFR and CSTR]}$$

$$30.34$$

$$\dot{m}_{\text{air}} = \frac{\dot{m}_{\text{oxygen}}}{0.232 \eta_{\text{transfer}}} \qquad 30.35$$

$$\dot{V}_{\text{air}} = \frac{\dot{m}_{\text{air}}}{\rho_{\text{air}}} \quad \text{[PFR and CSTR]} \qquad 30.36$$

$$m_{\text{air,m}^3/\text{kg BOD}} = \frac{\dot{V}_{\text{air}} \left(1000 \; \dfrac{\text{mg·m}^3}{\text{kg·L}}\right)}{Q_o (S_o - S)} \qquad 30.37$$

11. AERATION POWER AND COST

$$P_{\text{ideal,hp}} = -\left(\frac{k p_1 \dot{V}_1}{(k-1)\left(550 \; \dfrac{\text{ft-lbf}}{\text{hp-sec}}\right)}\right)\left(1 - \left(\frac{p_2}{p_1}\right)^{(k-1)/k}\right)$$

$$= -\left(\frac{k \dot{m} R_{\text{air}} T_1}{(k-1)\left(550 \; \dfrac{\text{ft-lbf}}{\text{hp-sec}}\right)}\right)\left(1 - \left(\frac{p_2}{p_1}\right)^{(k-1)/k}\right)$$

$$= -\left(\frac{c_p J \dot{m} T_1}{550 \; \dfrac{\text{ft-lbf}}{\text{hp-sec}}}\right)\left(1 - \left(\frac{p_2}{p_1}\right)^{(k-1)/k}\right)$$

$$30.38(b)$$

$$P_{\text{actual}} = \frac{P_{\text{ideal}}}{\eta_c} \qquad 30.39$$

$$\text{total cost} = C_{\text{kW-hr}} P_{\text{actual}} t \qquad 30.40$$

12. RETURN RATE/RECYCLE RATIO

$$R = \frac{Q_r}{Q_o} \qquad 30.41$$

$$X_o Q_o + X_r Q_r = X(Q_o + Q_r) \qquad 30.42$$

$$\frac{Q_r}{Q_o + Q_r} = \frac{X_r}{X} \qquad 30.43$$

$$\frac{Q_r}{Q_o + Q_r} = \frac{V_{\text{settled,mL/L}}}{1000 \; \dfrac{\text{mL}}{\text{L}}} \qquad 30.44$$

$$R = \frac{V_{\text{settled,mL/L}}}{1000 \; \dfrac{\text{mL}}{\text{L}} - V_{\text{settled,mL/L}}}$$

$$= \frac{1}{\dfrac{10^6}{(\text{SVI}_{\text{mL/g}})(\text{MLSS}_{\text{mg/L}})} - 1} \qquad 30.45$$

14. QUANTITIES OF SLUDGE

$$m_{\text{wet}} = V \rho_{\text{sludge}} = V (\text{SG}_{\text{sludge}}) \rho_{\text{water}} \qquad 30.46$$

$$\frac{1}{\text{SG}_{\text{sludge}}} = \frac{1 - s}{1} + \frac{s}{\text{SG}_{\text{solids}}}$$

$$= \frac{1 - s_{\text{fixed}} - s_{\text{volatile}}}{1} + \frac{s_{\text{fixed}}}{\text{SG}_{\text{fixed solids}}} + \frac{s_{\text{volatile}}}{\text{SG}_{\text{volatile solids}}} \qquad 30.47$$

$$V_{\text{sludge,wet}} = \frac{m_{\text{dried}}}{s \rho_{\text{sludge}}} \approx \frac{m_{\text{dried}}}{s \rho_{\text{water}}} \qquad 30.48$$

$$m_{\text{dried}} = (\Delta\text{SS})_{\text{mg/L}} Q_{o,\text{MGD}} \left(8.345 \, \frac{\text{lbm-L}}{\text{MG-mg}} \right) \quad 30.49$$

$$m_{\text{dried}} = K S_{o,\text{mg/L}} Q_{o,\text{MGD}} \left(8.345 \, \frac{\text{lbm-L}}{\text{MG-mg}} \right)$$
$$= Y_{\text{obs}}(S_{o,\text{mg/L}} - S) Q_{o,\text{MGD}} \left(8.345 \, \frac{\text{lbm-L}}{\text{MG-mg}} \right)$$
$$30.50$$

19. METHANE PRODUCTION

$$V_{\text{methane,ft}^3/\text{day}} = \left(5.61 \, \frac{\text{ft}^3}{\text{lbm}} \right) \begin{pmatrix} (E S_{o,\text{mg/L}} Q_{\text{MGD}}) \\ \times \left(8.345 \, \frac{\text{lbm-L}}{\text{MG-mg}} \right) \\ - 1.42 P_{x,\text{lbm/day}} \end{pmatrix}$$
$$30.51(b)$$

$$q = (\text{LHV}) V_{\text{fuel}} \quad 30.52$$

20. HEAT TRANSFER AND LOSS

$$q = m_{\text{sludge}} c_p (T_2 - T_1)$$
$$= V_{\text{sludge}} \rho_{\text{sludge}} c_p (T_2 - T_1) \quad \begin{bmatrix} \text{sensible} \\ \text{energy} \end{bmatrix} \quad 30.53$$

$$q = h A_{\text{surface}}(T_{\text{surface}} - T_{\text{air}}) \quad \begin{bmatrix} \text{convective} \\ \text{energy} \end{bmatrix} \quad 30.54$$

$$q = U A_{\text{surface}}(T_{\text{contents}} - T_{\text{air}}) \quad \begin{bmatrix} \text{total} \\ \text{energy} \end{bmatrix} \quad 30.55$$

$$q = \frac{k A (T_{\text{inner}} - T_{\text{outer}})}{L} \quad \begin{bmatrix} \text{conductive} \\ \text{energy} \end{bmatrix} \quad 30.56$$

21. SLUDGE DEWATERING

$$V_{\text{press}} \rho_{\text{filter cake}} s_{\text{filter cake}}$$
$$= V_{\text{sludge,per cycle}} s_{\text{sludge}} \rho_{\text{sludge}}$$
$$= V_{\text{sludge,per cycle}} s_{\text{sludge}} \rho_{\text{water}} \text{SG}_{\text{sludge}} \quad 30.57$$

$$G = \frac{\omega^2 r}{g} \quad [\text{centrifuge}] \quad 30.58$$

CERM Chapter 31
Municipal Solid Waste

> Chapter, section, equation, figure, and table numbers correspond to CERM. For additional study material, go to the corresponding chapter and section number in CERM.

3. LANDFILL CAPACITY

$$V_c = (\text{CF}) V_o \quad 31.1$$

$$\Delta V_{\text{day}} = \frac{NG(\text{LF})}{\gamma} = \frac{NG(\text{LF}) g_c}{\rho g} \quad 31.52(b)$$

$$\text{LF} = \frac{V_{\text{MSW}} + V_{\text{cover soil}}}{V_{\text{MSW}}} \quad 31.3$$

11. LANDFILL GAS

$$p_t = p_{\text{CH}_4} + p_{\text{CO}_2} + p_{\text{H}_2\text{O}} + p_{\text{N}_2} + p_{\text{other}} \quad 31.8$$
$$p_i = B_i p_t \quad 31.9$$

13. LEACHATE MIGRATION FROM LANDFILLS

$$Q = K i A \quad 31.13$$

$$i = \frac{dH}{dL} \quad 31.14$$

19. INCINERATION OF MUNICIPAL SOLID WASTE

$$\text{HRR} = \frac{(\text{fueling rate})(\text{HV})}{\text{total effective grate area}} \quad 31.24$$

CERM Chapter 32
Pollutants in the Environment

> Chapter, section, equation, figure, and table numbers correspond to CERM. For additional study material, go to the corresponding chapter and section number in CERM.

38. SMOKE

$$\text{optical density} = \log_{10} \left(\frac{1}{1 - \text{opacity}} \right) \quad 32.1$$

CERM Chapter 34
Environmental Remediation

> Chapter, section, equation, figure, and table numbers correspond to CERM. For additional study material, go to the corresponding chapter and section number in CERM.

16. CYCLONE SEPARATORS

$$S = \frac{v_{inlet}^2}{rg} \qquad 34.2$$

$$h = \frac{KBHv_{inlet}^2}{2gD^2} \qquad 34.3$$

18. ELECTROSTATIC PRECIPITATORS

$$\eta = 1 - K^{Ct} = 1 - e^{-wA_p/Q} \qquad 34.4$$

47. STRIPPING, AIR

$$p_A = H_A x_A \qquad 34.6$$

$$R = \frac{HG}{L} \qquad 34.7$$

Geotechnical

CERM Chapter 35
Soil Properties and Testing

Chapter, section, equation, figure, and table numbers correspond to CERM. For additional study material, go to the corresponding chapter and section number in CERM.

1. SOIL PARTICLE SIZE DISTRIBUTION

$$C_u = \frac{D_{60}}{D_{10}} \qquad 35.1$$

$$C_z = \frac{(D_{30})^2}{D_{10}D_{60}} \qquad 35.2$$

3. AASHTO SOIL CLASSIFICATION

$$I_g = (F_{200} - 35)\big(0.2 + 0.005(LL - 40)\big)$$
$$+ 0.01(F_{200} - 15)(PI - 10) \qquad 35.3$$

5. MASS-VOLUME RELATIONSHIPS

$$D_r = \frac{e_{\max} - e}{e_{\max} - e_{\min}} \quad \text{[relative density]} \qquad 35.16$$

Table 35.7 Soil Indexing Formulas

property		saturated sample (m_s, m_w, SG are known)	unsaturated sample (m_s, m_w, SG, V_t are known)	supplementary formulas relating measured and computed factors			
volume components							
V_s	volume of solids		$\frac{m_s}{(SG)\rho_w}$	$V_t - (V_g + V_w)$	$V_t(1-n)$	$\frac{V_t}{1+e}$	$\frac{V_v}{e}$
V_w	volume of water		$\frac{m_w}{\rho_w^*}$	$V_v - V_g$	SV_v	$\frac{SV_t e}{1+e}$	$SV_s e$
V_g	volume of gas or air	zero	$V_t - (V_s + V_w)$	$V_v - V_w$	$(1-S)V_v$	$\frac{(1-S)V_t e}{1+e}$	$(1-S)V_s e$
V_v	volume of voids	$\frac{m_w}{\rho_w^*}$	$V_t - \frac{m_s}{(SG)\rho_w}$	$V_t - V_s$	$\frac{V_s n}{1-n}$	$\frac{V_t e}{1+e}$	$V_s e$
V_t	total volume of sample	$V_s + V_w$	measured $(V_g + V_w + V_s)$	$V_s + V_g + V_w$	$\frac{V_s}{1-n}$	$V_s(1+e)$	$\frac{V_v(1+e)}{e}$
n	porosity		$\frac{V_v}{V_t}$	$1 - \frac{V_s}{V_t}$	$1 - \frac{m_s}{(SG)V_t\rho_w}$	$\frac{e}{1+e}$	
e	void ratio		$\frac{V_v}{V_s}$	$\frac{V_t}{V_s} - 1$	$\frac{(SG)V_t\rho_w}{m_s} - 1$	$\frac{m_w(SG)}{m_s S}$	$\frac{n}{1-n}\left|w\left(\frac{SG}{S}\right)\right.$

(continued)

Table 35.7 *Soil Indexing Formulas (continued)*

property	saturated sample (m_s, m_w, SG are known)	unsaturated sample (m_s, m_w, SG, V_t are known)	supplementary formulas relating measured and computed factors				
mass for specific sample							
m_s mass of solids		measured	$\dfrac{m_t}{1+w}$	$(SG)V_t\rho_w(1-n)$	$\dfrac{m_w(SG)}{eS}$	$V_s(SG)\rho_w$	
m_w mass of water		measured	wm_s	$S\rho_w V_v$	$\dfrac{em_s S}{SG}$	$V_t\rho_d w$	
m_t total mass of sample		$m_s + m_w$	$m_s(1+w)$				
mass for sample of unit volume (density)							
ρ_d dry density	$\dfrac{m_s}{V_s+V_w}$	$\dfrac{m_s}{V_g+V_w+V_s}$	$\dfrac{m_t}{V_t(1+w)}$	$\dfrac{(SG)\rho_w}{1+e}$	$\dfrac{(SG)\rho_w}{1+\dfrac{w(SG)}{S}}$	$\dfrac{\rho}{1+w}$	
ρ wet density (density with moisture)	$\dfrac{m_s+m_w}{V_s+V_w}$	$\dfrac{m_s+m_w}{V_t}$	$\dfrac{m_t}{V_t}$	$\dfrac{(SG+Se)\rho_w}{1+e}$	$\dfrac{(1+w)\rho_w}{\dfrac{w}{S}+\dfrac{1}{SG}}$	$\rho_d(1+w)$	
ρ_{sat} saturated density	$\dfrac{m_s+m_w}{V_s+V_w}$	$\dfrac{m_s+V_v\rho_w}{V_t}$	$\dfrac{m_s}{V_t}+\left(\dfrac{e}{1+e}\right)\rho_w$	$\dfrac{(SG+e)\rho_w}{1+e}$	$\dfrac{(1+w)\rho_w}{w+\dfrac{1}{SG}}$		
ρ_b buoyant (submerged) density		$\rho_{\text{sat}}-\rho_w^*$	$\dfrac{m_s}{V_t}-\left(\dfrac{1}{1+e}\right)\rho_w^*$	$\left(\dfrac{SG+e}{1+e}-1\right)\rho_w^*$	$\left(\dfrac{1-\dfrac{1}{SG}}{w+\dfrac{1}{SG}}\right)\rho_w^*$		
combined relations							
w water content		$\dfrac{m_w}{m_s}$	$\dfrac{m_t}{m_s}-1$	$\dfrac{Se}{SG}$	$S\left(\dfrac{\rho_w^*}{\rho_d}-\dfrac{1}{SG}\right)$		
S degree of saturation	100%	$\dfrac{V_w}{V_v}$	$\dfrac{m_w}{V_v\rho_w^*}$	$\dfrac{w(SG)}{e}$	$\dfrac{w}{\dfrac{\rho_w^*}{\rho_d}-\dfrac{1}{SG}}$		
SG specific gravity of solids		$\dfrac{m_s}{V_s\rho_w}$	$\dfrac{Se}{w}$				

ρ_w is the density of water. Where noted with an asterisk (*), use the actual density of water at the recorded temperature. In other cases, use 62.4 lbm/ft^3 or 1000 kg/m^3.

7. EFFECTIVE STRESS

$$\sigma' = \sigma - u \qquad 35.17$$

10. CONE PENETROMETER TEST

$$f_R = \frac{q_s}{q_c} \times 100\% \qquad 35.18$$

11. PROCTOR TEST

$$\text{RC} = \frac{\rho_d}{\rho_d^*} \times 100\% \qquad 35.19$$

12. MODIFIED PROCTOR TEST

$$\rho_z = \frac{m_s}{V_w + V_s} \qquad 35.20$$

$$\rho_z = \frac{\rho_w}{w + \dfrac{1}{\text{SG}}} \qquad 35.21$$

$$\rho_s = (\text{SG})\rho_w \qquad 35.22$$

14. ATTERBERG LIMIT TESTS

$$\text{PI} = \text{LL} - \text{PL} \qquad 35.23$$

$$\text{SI} = \text{PL} - \text{SL} \qquad 35.24$$

$$\text{LI} = \frac{w - \text{PL}}{\text{PI}} \qquad 35.25$$

15. PERMEABILITY TESTS

$$Q = \text{v}A_{\text{gross}} \qquad 35.26$$

$$\text{v} = Ki \qquad 35.27$$

$$K_{\text{cm/s}} \approx C(D_{10,\text{mm}})^2 \qquad 35.28$$

$$K = \frac{VL}{hAt} \qquad \text{[constant head]} \qquad 35.29$$

$$K = \left(\frac{A'L}{At}\right)\ln\left(\frac{h_i}{h_f}\right) \qquad \text{[falling head]} \qquad 35.30$$

Figure 35.8 Permeameters

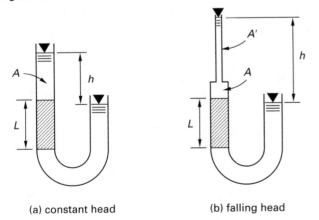

(a) constant head   (b) falling head

$$F = \frac{11r}{2} \qquad \text{[auger hole]} \qquad 35.31$$

$$K = \left(\frac{\pi r^2}{Ft}\right)\ln\left(\frac{h_i}{h_f}\right)$$

$$= \left(\frac{2\pi r}{11t}\right)\ln\left(\frac{h_i}{h_f}\right) \qquad \text{[auger hole]} \qquad 35.32$$

16. CONSOLIDATION TESTS

$$\text{OCR} = \frac{p'_{\max}}{p'_o} \qquad 35.33$$

$$C_c = -\frac{e_1 - e_2}{\log_{10}\left(\dfrac{p_1}{p_2}\right)} \qquad 35.34$$

$$C_{\epsilon c} = \frac{C_c}{1 + e_0} \qquad 35.35$$

$$C_c \approx (0.009)(\text{LL} - 10) \qquad 35.36$$

Figure 35.10 e-log p Curve

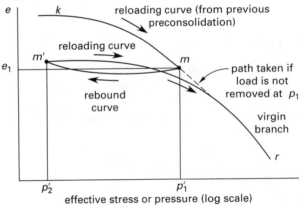

17. DIRECT SHEAR TEST

$$S = \tau = c + \sigma \tan\phi \qquad 35.37$$

18. TRIAXIAL STRESS TEST

$$\sigma_\alpha = \tfrac{1}{2}(\sigma_A + \sigma_R)$$
$$\quad + \tfrac{1}{2}(\sigma_A - \sigma_R)\cos 2\alpha \qquad 35.38$$

$$\tau_\alpha = \tfrac{1}{2}(\sigma_A - \sigma_R)\sin 2\alpha \qquad 35.39$$

$$\frac{\sigma_1}{\sigma_3} = \frac{1 + \sin\phi}{1 - \sin\phi} \quad [c = 0] \qquad 35.40$$

$$\alpha = 45° + \tfrac{1}{2}\phi \qquad 35.41$$

$$S_{us} = c = \frac{\sigma_D}{2} \qquad 35.42$$

$$s = c' + \sigma' \tan\phi' \qquad 35.43$$

20. UNCONFINED COMPRESSIVE STRENGTH TEST

$$S_{\mathrm{uc}} = \frac{P}{A} \qquad 35.44$$

$$s_u = \frac{S_{\mathrm{uc}}}{2} \qquad 35.45$$

21. SENSITIVITY

$$S_t = \frac{S_{\mathrm{undisturbed}}}{S_{\mathrm{remolded}}} \qquad 35.46$$

22. CALIFORNIA BEARING RATIO TEST

$$\mathrm{CBR} = \frac{\mathrm{actual\ load}}{\mathrm{standard\ load}} \times 100 \qquad 35.47$$

CERM Chapter 36
Shallow Foundations

Chapter, section, equation, figure, and table numbers correspond to CERM. For additional study material, go to the corresponding chapter and section number in CERM.

5. GENERAL BEARING CAPACITY EQUATION

$$q_{\mathrm{ult}} = \tfrac{1}{2}\gamma B N_\gamma + c N_c + (p_q + \gamma D_f)N_q \qquad 36.1(b)$$

Table 36.2 *Terzaghi Bearing Capacity Factors for General Shear[a]*

ϕ	N_c	N_q	N_γ
0	5.7	1.0	0.0
5	7.3	1.6	0.5
10	9.6	2.7	1.2
15	12.9	4.4	2.5
20	17.7	7.4	5.0
25	25.1	12.7	9.7
30	37.2	22.5	19.7
34	52.6	36.5	35.0
35	57.8	41.4	42.4
40	95.7	81.3	100.4
45	172.3	173.3	297.5
48	258.3	287.9	780.1
50	347.5	415.1	1153.2

[a]Curvilinear interpolation may be used. Do not use linear interpolation.

Table 36.3 *Meyerhof and Vesic Bearing Capacity Factors for General Shear[a]*

ϕ	N_c	N_q	N_γ	$N_\gamma{}^b$
0	5.14	1.0	0.0	0.0
5	6.5	1.6	0.07	0.5
10	8.3	2.5	0.37	1.2
15	11.0	3.9	1.1	2.6
20	14.8	6.4	2.9	5.4
25	20.7	10.7	6.8	10.8
30	30.1	18.4	15.7	22.4
32	35.5	23.2	22.0	30.2
34	42.2	29.4	31.2	41.1
36	50.6	37.7	44.4	56.3
38	61.4	48.9	64.1	78.0
40	75.3	64.2	93.7	109.4
42	93.7	85.4	139.3	155.6
44	118.4	115.3	211.4	224.6
46	152.1	158.5	328.7	330.4
48	199.3	222.3	526.5	496.0
50	266.9	319.1	873.9	762.9

[a]Curvilinear interpolation may be used. Do not use linear interpolation.
[b]As predicted by the Vesic equation, $N_\gamma = 2(N_q + 1)\tan\phi$.

Table 36.4 *N_c Bearing Capacity Factor Multipliers for Various Values of B/L*

B/L	multiplier
1 (square)	1.25
0.5	1.12
0.2	1.05
0.0	1.00
1 (circular)	1.20

Table 36.5 N_γ Multipliers for Various Values of B/L

B/L	multiplier
1 (square)	0.85
0.5	0.90
0.2	0.95
0.0	1.00
1 (circular)	0.70

$$q_{\text{net}} = q_{\text{ult}} - \gamma D_f \qquad 36.3(b)$$

$$q_a = \frac{q_{\text{net}}}{F} \qquad 36.4$$

6. BEARING CAPACITY OF CLAY

$$S_u = c = \frac{S_{\text{uc}}}{2} \qquad 36.5$$

$$q_{\text{ult}} = cN_c + \gamma D_f \qquad 36.6(b)$$

$$q_{\text{net}} = q_{\text{ult}} - \gamma D_f = cN_c \qquad 36.7(b)$$

$$q_a = \frac{q_{\text{net}}}{F} \qquad 36.8$$

7. BEARING CAPACITY OF SAND

$$q_{\text{ult}} = \tfrac{1}{2}B\gamma N_\gamma + (p_q + \gamma D_f)N_q \qquad 36.9(b)$$

$$q_{\text{net}} = q_{\text{ult}} - \gamma D_f$$
$$= \tfrac{1}{2}B\gamma N_\gamma + \gamma D_f(N_q - 1) \qquad 36.10(b)$$

$$q_a = \frac{q_{\text{net}}}{F}$$
$$= \left(\frac{B}{F}\right)\left(\tfrac{1}{2}\gamma N_\gamma + \gamma(N_q - 1)\left(\frac{D_f}{B}\right)\right) \qquad 36.11(b)$$

$$q_a = 0.11 C_n N \quad \begin{bmatrix} \text{in tons/ft}^2,\ B > 2\text{--}4\ \text{ft,} \\ N \le 50 \end{bmatrix} \qquad 36.12$$

9. EFFECTS OF WATER TABLE ON FOOTING DESIGN

$$q_{\text{ult}} = \tfrac{1}{2}\gamma_b B N_\gamma + \gamma_d D_f N_q$$
$$\begin{bmatrix} \text{sand; water table at} \\ \text{base of footing; } c = 0 \end{bmatrix} \qquad 36.13(b)$$

$$q_{\text{ult}} = \tfrac{1}{2}\gamma_b B N_\gamma + \gamma_b D_f N_q$$
$$\begin{bmatrix} \text{sand; water table} \\ \text{at surface; } c = 0 \end{bmatrix} \qquad 36.14(b)$$

$$\left(\gamma_d D_w + \left(\gamma_d - 62.4\ \frac{\text{lbf}}{\text{ft}^3}\right)(D_f - D_w)\right)N_q$$
$$= \left(\gamma_d D_f + \left(62.4\ \frac{\text{lbf}}{\text{ft}^3}\right)(D_w - D_f)\right)N_q$$
$$\begin{bmatrix} \text{sand; water table between} \\ \text{base and surface; } c = 0 \end{bmatrix} \qquad 36.15(b)$$

$$q_{\text{ult}} = \tfrac{1}{2}\gamma_d B N_\gamma + \gamma_d D_f N_q$$
$$\begin{bmatrix} \text{sand; } D_w > D_f + B; \\ c = 0 \end{bmatrix} \qquad 36.16(b)$$

10. ECCENTRIC LOADS ON RECTANGULAR FOOTINGS

$$\epsilon_B = \frac{M_B}{P}; \quad \epsilon_L = \frac{M_L}{P} \qquad 36.17$$

$$L' = L - 2\epsilon_L; \quad B' = B - 2\epsilon_B \qquad 36.18$$

$$A' = L'B' \qquad 36.19$$

Figure 36.6 Footing with Overturning Moment

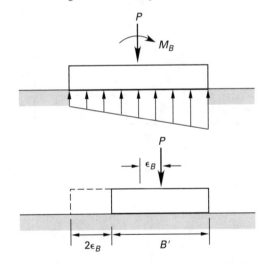

$$p_{\text{max}},\ p_{\text{min}} = \left(\frac{P}{BL}\right)\left(1 \pm \frac{6\epsilon}{B}\right) \qquad 36.20$$

12. RAFTS ON CLAY

$$F = \frac{cN_c}{\dfrac{\text{total load}}{\text{raft area}} - \gamma D_f} \qquad 36.21(b)$$

13. RAFTS ON SAND

$$q_a = 0.22 C_n N \quad [\text{in tons/ft}^2] \qquad 36.22$$

$$p = \frac{\text{total load}}{\text{raft area}} - \gamma D_f \qquad 36.23(b)$$

CERM Chapter 37
Rigid Retaining Walls

> Chapter, section, equation, figure, and table numbers correspond to CERM. For additional study material, go to the corresponding chapter and section number in CERM.

3. EARTH PRESSURE

Figure 37.2 *Active and Passive Earth Pressure*

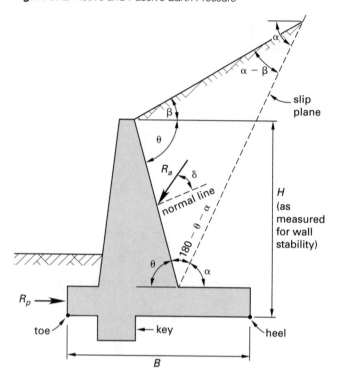

$$\alpha = 45° + \frac{\phi}{2} \quad \text{[Rankine]} \qquad 37.1$$

$$\alpha = \phi + \arctan$$
$$\times \left(\frac{-\tan \phi + \sqrt{\tan \phi (\tan \phi + \cot \phi)(1 + \tan \delta \cot \phi)}}{1 + \tan \delta (\tan \phi + \cot \phi)} \right)$$
$$\text{[Coulomb]} \qquad 37.2$$

4. VERTICAL SOIL PRESSURE

$$p_v = \gamma H \qquad 37.3(b)$$

5. ACTIVE EARTH PRESSURE

$$p_a = p_v k_a - 2c\sqrt{k_a} \qquad 37.4$$

$$k_a = \frac{\sin^2(\theta + \phi)}{\sin^2 \theta \sin(\theta - \delta) \left(1 + \sqrt{\dfrac{\sin(\phi + \delta) \sin(\phi - \beta)}{\sin(\theta - \delta) \sin(\theta + \beta)}} \right)^2}$$
$$\text{[Coulomb]} \qquad 37.5$$

$$k_a = \cos \beta \left(\frac{\cos \beta - \sqrt{\cos^2 \beta - \cos^2 \phi}}{\cos \beta + \sqrt{\cos^2 \beta - \cos^2 \phi}} \right) \quad \text{[Rankine]}$$
$$37.6$$

$$k_a = \frac{1}{k_p} = \tan^2 \left(45° - \frac{\phi}{2} \right)$$
$$= \frac{1 - \sin \phi}{1 + \sin \phi} \quad \left[\begin{array}{l} \text{Rankine: horizontal} \\ \text{backfill; vertical face} \end{array} \right] \qquad 37.7$$

$$p_a = p_v - 2c \quad [\phi = 0] \qquad 37.8$$

$$p_a = k_a p_v \quad [c = 0] \qquad 37.9$$

$$R_a = \tfrac{1}{2} p_a H = \tfrac{1}{2} k_a \gamma H^2 \qquad 37.10(b)$$

$$\theta_R = 90° \text{ from the wall} \qquad \text{[Rankine]} \qquad 37.11$$

$$\theta_R = 90° - \delta \text{ from the wall} \quad \text{[Coulomb]} \qquad 37.12$$

6. PASSIVE EARTH PRESSURE

$$p_p = p_v k_p + 2c\sqrt{k_p} \qquad 37.13$$

$$k_p = \frac{\sin^2(\theta - \phi)}{\sin^2 \theta \sin(\theta + \delta) \left(1 - \sqrt{\dfrac{\sin(\phi + \delta) \sin(\phi + \beta)}{\sin(\theta + \delta) \sin(\theta + \beta)}} \right)^2}$$
$$\text{[Coulomb]} \qquad 37.14$$

$$k_p = \cos \beta \left(\frac{\cos \beta + \sqrt{\cos^2 \beta - \cos^2 \phi}}{\cos \beta - \sqrt{\cos^2 \beta - \cos^2 \phi}} \right) \quad \text{[Rankine]} \qquad 37.15$$

$$k_p = \frac{1}{k_a} = \tan^2 \left(45° + \frac{\phi}{2} \right)$$
$$= \frac{1 + \sin \phi}{1 - \sin \phi} \quad \left[\begin{array}{l} \text{Rankine: horizontal} \\ \text{backfill; vertical face} \end{array} \right] \qquad 37.16$$

$$p_p = p_v + 2c \quad [\phi = 0] \qquad 37.17$$

$$p_p = k_p p_v \quad [c = 0] \qquad 37.18$$

$$R_p = \tfrac{1}{2} p_p H = \tfrac{1}{2} k_p \gamma H^2 \qquad 37.19(b)$$

7. AT-REST SOIL PRESSURE

$$p_o = k_o p_v \qquad 37.20$$
$$k_o \approx 1 - \sin\phi \qquad 37.21$$
$$R_o = \tfrac{1}{2} k_o \gamma H^2 \qquad 37.22(b)$$

9. SURCHARGE LOADING

$$p_q = k_a q \qquad 37.27$$
$$R_q = k_a q H \times (\text{wall width}) \qquad 37.28$$
$$p_q = \frac{1.77 V_q m^2 n^2}{H^2 (m^2 + n^2)^3} \quad [m > 0.4] \qquad 37.29$$
$$p_q = \frac{0.28 V_q n^2}{H^2 (0.16 + n^2)^3} \quad [m \le 0.4] \qquad 37.30$$
$$R_q \approx \frac{0.78 V_q}{H} \quad [m = 0.4] \qquad 37.31$$
$$R_q \approx \frac{0.60 V_q}{H} \quad [m = 0.5] \qquad 37.32$$
$$R_q \approx \frac{0.46 V_q}{H} \quad [m = 0.6] \qquad 37.33$$
$$m = \frac{x}{H} \qquad 37.34$$
$$n = \frac{y}{H} \qquad 37.35$$

$$p_q = \frac{4 L_q m^2 n}{\pi H (m^2 + n^2)^2} \quad [m > 0.4] \qquad 37.36$$
$$R_q = \frac{0.64 L_q}{m^2 + 1} \qquad 37.37$$
$$p_q = \frac{0.203 L_q n}{H(0.16 + n^2)^2} \quad [m \le 0.4] \qquad 37.38$$
$$R_q = 0.55 L_q \qquad 37.39$$

Figure 37.3 *Surcharges*

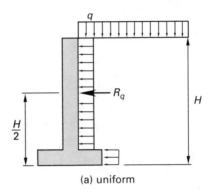

(a) uniform

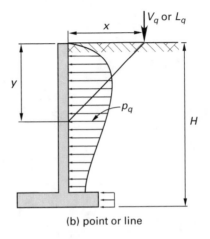

(b) point or line

10. EFFECTIVE STRESS

$$\mu = \gamma_w h \qquad 37.40(b)$$
$$\gamma_{\text{sat}} = \gamma_{\text{dry}} + n\gamma_w$$
$$= \gamma_{\text{dry}} + \left(\frac{e}{1+e}\right)\gamma_w \qquad 37.41(b)$$
$$p_v = \gamma_{\text{sat}} H - \gamma_w h \qquad 37.42(b)$$
$$p_h = \gamma_w h + k_a(\gamma_{\text{sat}} H - \gamma_w h)$$
$$= k_a \gamma_{\text{sat}} H + (1 - k_a)\gamma_w h \qquad 37.43(b)$$
$$\gamma_{\text{eq}} = (1 - k_a)\gamma_w \qquad 37.44(b)$$

11. CANTILEVER RETAINING WALLS: ANALYSIS

$$W_i = \gamma_i A_i \qquad 37.45(b)$$
$$M_{\text{toe}} = \sum W_i x_i - R_{a,h} y_a + R_{a,v} x_a \qquad 37.46$$

$$x_R = \frac{M_{\text{toe}}}{R_{a,v} + \sum W_i} \qquad 37.47$$

$$\epsilon = \left| \frac{B}{2} - x_R \right| \qquad 37.48$$

$$F_{\text{OT}} = \frac{M_{\text{resisting}}}{M_{\text{overturning}}}$$
$$= \frac{\sum W_i x_i + R_{a,v} x_{a,v}}{R_{a,h} y_{a,h}} \qquad 37.49$$

$$p_{v,\text{max}},\ p_{v,\text{min}} = \left(\frac{\sum W_i + R_{a,v}}{B} \right)$$
$$\times \left(1 \pm \left(\frac{6\epsilon}{B} \right) \right) \qquad 37.50$$

$$R_{\text{SL}} = \left(\sum W_i + R_{a,v} \right) \tan \delta + c_A B \qquad 37.52$$

$$F_{\text{SL}} = \frac{R_{\text{SL}}}{R_{a,h}} \qquad 37.53$$

Figure 37.4 *Elements Contributing to Vertical Force[a] (step 3)*

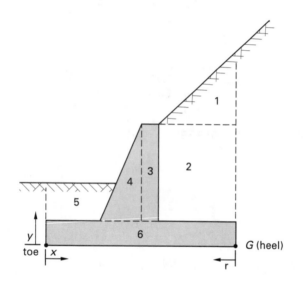

[a]Some retaining walls may not have all elements.

Figure 37.5 *Resultant Distribution on the Base (step 7)*

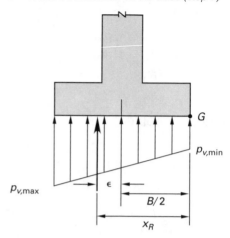

12. RETAINING WALLS: DESIGN

$$R_{a,h} \left(\frac{H}{3} \right) \approx (\text{soil weight}) \left(\frac{L}{2} \right) \qquad 37.54$$

$$B \approx \tfrac{3}{2} L \qquad 37.55$$

$$8 < \frac{H}{t_{\text{stem}}} < 12 \qquad 37.56$$

$$10 < \frac{H}{t_{\text{base}}} < 14 \qquad 37.57$$

CERM Chapter 38
Piles and Deep Foundations

Chapter, section, equation, figure, and table numbers correspond to CERM. For additional study material, go to the corresponding chapter and section number in CERM.

1. INTRODUCTION

$$Q_{\text{ult}} = Q_p + Q_f \qquad 38.1$$

$$Q_a = \frac{Q_{\text{ult}}}{F} \qquad 38.2$$

2. PILE CAPACITY FROM DRIVING DATA

$$Q_{a,\text{lbf}} = \frac{2W_{\text{hammer,lbf}} H_{\text{fall,ft}}}{S_{\text{in}} + 1} \quad [\text{drop hammer}] \qquad 38.3$$

$$Q_{a,\text{lbf}} = \frac{2W_{\text{hammer,lbf}} H_{\text{fall,ft}}}{S_{\text{in}} + 0.1}$$
$$\left[\begin{array}{l} \text{single-acting steam hammer;} \\ \text{driven weight} < \text{striking weight} \end{array} \right] \qquad 38.4$$

$$Q_{a,\text{lbf}} = \frac{2W_{\text{hammer,lbf}} H_{\text{fall,ft}}}{S_{\text{in}} + 0.1 \left(\dfrac{W_{\text{driven}}}{W_{\text{hammer}}} \right)}$$
$$\left[\begin{array}{l} \text{single-acting steam hammer;} \\ \text{driven weight} > \text{striking weight} \end{array} \right] \qquad 38.5$$

$$Q_{a,\text{lbf}} = \frac{2E_{\text{ft-lbf}}}{S_{\text{in}} + 0.1}$$

$$\left[\begin{array}{l}\text{double-acting steam hammer;} \\ \text{driven weight} < \text{striking weight}\end{array}\right] \quad 38.6$$

$$Q_{a,\text{lbf}} = \frac{2E_{\text{ft-lbf}}}{S_{\text{in}} + 0.1\left(\dfrac{W_{\text{driven}}}{W_{\text{hammer}}}\right)}$$

$$\left[\begin{array}{l}\text{double-acting steam hammer;} \\ \text{driven weight} > \text{striking weight}\end{array}\right] \quad 38.7$$

3. THEORETICAL POINT-BEARING CAPACITY

$$Q_p = A_p\left(\tfrac{1}{2}\gamma B N_\gamma + c N_c + \gamma D_f N_q\right) \qquad 38.8(b)$$

$$Q_p = A_p \gamma D N_q \quad [\text{cohesionless; } D \leq D_c] \qquad 38.9(b)$$

$$Q_p = A_p c N_c \approx 9 A_p c \quad [\text{cohesive}] \qquad 38.10$$

4. THEORETICAL SKIN-FRICTION CAPACITY

$$\begin{aligned} Q_f &= A_s f_s \\ &= p f_s L_e = p f_s (L - \text{seasonal variation}) \end{aligned} \qquad 38.11$$

$$Q_f = p\sum f_{s,i} L_{e,i} \qquad 38.12$$

$$f_s = c_A + \sigma_h \tan\delta \qquad 38.13$$

$$f_s = \alpha c \qquad 38.14$$

$$\sigma_h = k_s \sigma'_v = k_s(\gamma D - \mu) \qquad 38.15(b)$$

$$\mu = \gamma_w h \qquad 38.16(b)$$

$$Q_f = p k_s \tan\delta \sum L_i \sigma' \quad [\text{cohesionless}] \qquad 38.17$$

$$Q_f = p\sum c_A L_i \quad [\text{cohesive}] \qquad 38.18$$

$$Q_f = p\beta\sigma' L \quad [\text{cohesive}] \qquad 38.19$$

7. CAPACITY OF PILE GROUPS

$$Q_s = 2(b+w)L_e c_1 \qquad 38.20$$

$$Q_p = 9c_2 bw \qquad 38.21$$

$$Q_{\text{ult}} = Q_s + Q_p \qquad 38.22$$

$$Q_a = \frac{Q_{\text{ult}}}{F} \qquad 38.23$$

$$\eta_G = \frac{\text{group capacity}}{\sum \text{individual capacities}} \qquad 38.24$$

CERM Chapter 39
Excavations

> Chapter, section, equation, figure, and table numbers correspond to CERM. For additional study material, go to the corresponding chapter and section number in CERM.

3. BRACED CUTS IN SAND

$$p_{\max} = 0.65 k_a \gamma H \qquad 39.1(b)$$

Figure 39.2 *Cuts in Sand*

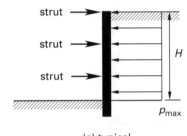

(a) typical

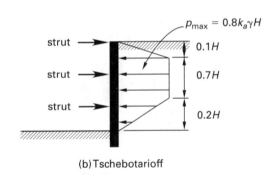

(b) Tschebotarioff

4. BRACED CUTS IN STIFF CLAY

$$0.2\gamma H \leq p_{\max} \leq 0.4\gamma H \qquad 39.2(b)$$

$$p_{\max} = k_a \gamma H \qquad 39.3(b)$$

$$k_a = 1 - \frac{4c}{\gamma H} \qquad 39.4(b)$$

Figure 39.3 *Cuts in Stiff Clay*

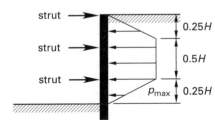

5. BRACED CUTS IN SOFT CLAY

$$p_{max} = \gamma H - 4c \qquad \text{39.5(b)}$$

Figure 39.4 *Cuts in Soft Clay*

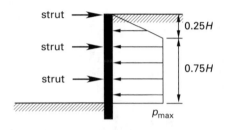

7. ANALYSIS/DESIGN OF BRACED EXCAVATIONS

$$S = \frac{M_{max,\text{sheet piling}}}{F_b} \qquad \text{39.6}$$

8. STABILITY OF BRACED EXCAVATIONS IN CLAY

$$H_c = \frac{5.7c}{\gamma - \sqrt{2}\left(\dfrac{c}{B}\right)} \quad [H < B] \qquad \text{39.7(b)}$$

$$H_c = \frac{N_c c}{\gamma} \quad [H > B] \qquad \text{39.8(b)}$$

$$F = \frac{N_c c}{\gamma H + q} \qquad \text{39.9(b)}$$

9. STABILITY OF BRACED EXCAVATIONS IN SAND

$$F = 2N_\gamma k_a \tan \phi \qquad \text{39.10}$$

$$F = 2N_\gamma \left(\frac{\gamma_{\text{submerged}}}{\gamma_{\text{drained}}}\right) k_a \tan \phi \qquad \text{39.11(b)}$$

11. ANALYSIS/DESIGN OF FLEXIBLE BULKHEADS

$$y = \frac{k_p D^2 - k_a (H + D)^2}{(k_p - k_a)(H + 2D)} \qquad \text{39.12}$$

Figure 39.5 *Shear on a Cantilever Wall in Uniform Granular Soil*

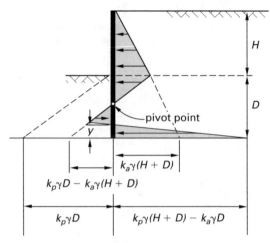

13. ANALYSIS/DESIGN OF TIED BULKHEADS

$$f = \frac{M}{S} \qquad \text{39.13}$$

$$S = \frac{M}{F_b} \qquad \text{39.14}$$

CERM Chapter 40
Special Soil Topics

> Chapter, section, equation, figure, and table numbers correspond to CERM. For additional study material, go to the corresponding chapter and section number in CERM.

1. PRESSURE FROM APPLIED LOADS: BOUSSINESQ'S EQUATION

$$\Delta p_v = \frac{3h^3 P}{2\pi z^5}$$

$$= \left(\frac{3P}{2\pi h^2}\right)\left(\frac{1}{1 + \left(\dfrac{r}{h}\right)^2}\right)^{5/2} \quad [h > 2B] \quad \text{40.1}$$

Figure 40.1 *Pressure at a Point*

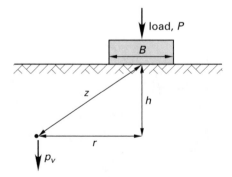

2. PRESSURE FROM APPLIED LOADS: ZONE OF INFLUENCE

$$A = \left(B + 2(h \cot 60°)\right)\left(L + 2(h \cot 60°)\right) \quad 40.2$$

$$A = (B + h)(L + h) \quad 40.3$$

$$\Delta p_v = \frac{P}{A} \quad 40.4$$

3. PRESSURE FROM APPLIED LOADS: INFLUENCE CHART

$$\Delta p_v = (\text{influence value})(\text{no. of squares}) \\ \times (\text{applied pressure}) \quad 40.5$$

5. CLAY CONDITION

Figure 40.4 *Consolidation Curve for Clay*

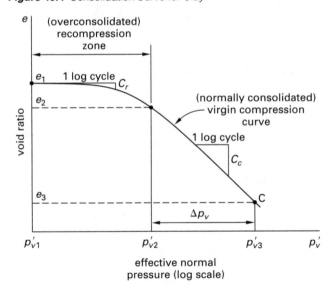

6. CONSOLIDATION PARAMETERS

$$C_r = \frac{-(e_1 - e_2)}{\log_{10}\left(\dfrac{p_1}{p_2}\right)} = \frac{e_2 - e_1}{\log_{10}\left(\dfrac{p_1}{p_2}\right)} \quad 40.6$$

$$C_c = \frac{-(e_2 - e_3)}{\log_{10}\left(\dfrac{p_2}{p_3}\right)} = \frac{e_3 - e_2}{\log_{10}\left(\dfrac{p_2}{p_3}\right)}$$

$$= \frac{\Delta e}{\log_{10}\left(\dfrac{p_2}{p_2 + \Delta p_v}\right)} \quad 40.7$$

$$C_c \approx 1.15(e_o - 0.35) \quad [\text{clays}] \quad 40.8$$

$$C_c \approx 0.009(\text{LL} - 10) \quad \begin{bmatrix}\text{normally} \\ \text{consolidated clays}\end{bmatrix} \quad 40.9$$

$$C_c \approx 0.0155w \quad [\text{organic soils}] \quad 40.10$$

$$C_c \approx (1 + e_o)\left(0.1 + 0.006(w_n - 25)\right) \\ [\text{varved clays}] \quad 40.11$$

$$\text{CR} = \frac{C_c}{1 + e_o} \quad 40.12$$

$$\text{RR} = \frac{C_r}{1 + e_o} \quad 40.13$$

$$e_o = w_o(\text{SG}) \quad [\text{saturated}] \quad 40.14$$

7. PRIMARY CONSOLIDATION

$$S_{\text{primary}} = \frac{H \Delta e}{1 + e_o}$$

$$= \frac{H C_r \log_{10}\left(\dfrac{p_o' + \Delta p_v'}{p_o'}\right)}{1 + e_o}$$

$$= H(\text{RR}) \log_{10}\left(\frac{p_o' + \Delta p_v'}{p_o'}\right) \quad [\text{overconsolidated}] \quad 40.15$$

$$S_{\text{primary}} = \frac{H \Delta e}{1 + e_o}$$

$$= \frac{H C_c \log_{10}\left(\dfrac{p_o' + \Delta p_v'}{p_o'}\right)}{1 + e_o}$$

$$= H(\text{CR}) \log_{10}\left(\frac{p_o' + \Delta p_v'}{p_o'}\right) \quad \begin{bmatrix}\text{normally} \\ \text{consolidated}\end{bmatrix} \quad 40.16$$

$$p_v = \gamma_{\text{layer}} H \quad [\text{above GWT}] \quad 40.17$$

$$p_v' = \gamma_{\text{layer}} H - \gamma_{\text{water}} h$$

$$= p_v - \mu \quad [\text{below GWT}] \quad 40.18$$

$$\mu = \gamma_{\text{water}} h \quad 40.19$$

8. PRIMARY CONSOLIDATION RATE

$$t = \frac{T_v H_d^2}{C_v} \quad 40.20$$

$$C_v = \frac{K(1 + e_o)}{a_v \gamma_{\text{water}}} \quad 40.21$$

$$a_v = \frac{-(e_2 - e_1)}{p_2' - p_1'} \quad 40.22$$

$$T_v = \tfrac{1}{4}\pi U_z^2 \quad [U_z < 0.60] \quad 40.23$$

Figure 40.5 *Consolidation Parameters*

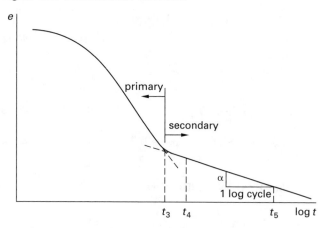

9. SECONDARY CONSOLIDATION

$$\alpha = \frac{-(e_5 - e_4)}{\log\left(\dfrac{t_5}{t_4}\right)} \qquad 40.24$$

$$C_\alpha = \frac{\alpha}{1 + e_o} \qquad 40.25$$

$$S_{\text{secondary}} = C_\alpha H \log_{10}\left(\frac{t_5}{t_4}\right) \qquad 40.26$$

10. SLOPE STABILITY IN SATURATED CLAY

$$d = \frac{D}{H} \qquad 40.27$$

$$F_{\text{cohesive}} = \frac{N_o c}{\gamma_{\text{eff}} H} \qquad 40.28$$

$$\gamma_{\text{eff}} = \gamma_{\text{saturated}} - \gamma_{\text{water}} \qquad 40.29$$

Figure 40.6 *Taylor Slope Stability*
$$(\phi = 0°)$$

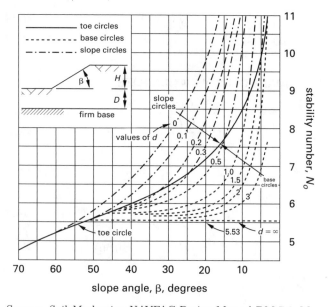

Source: *Soil Mechanics*, NAVFAC Design Manual DM-7.1, May 1982

11. LOADS ON BURIED PIPES

$$w = C\gamma B^2 \quad \begin{bmatrix} \text{Marston's formula;} \\ \text{rigid pipe} \end{bmatrix} \qquad 40.30(b)$$

$$p = \frac{w}{B} \qquad 40.31$$

$$w = C\gamma BD \quad [\text{flexible pipe}] \qquad 40.32(b)$$

$$w = C_p \gamma D^2 \quad [\text{broad fill}] \qquad 40.33(b)$$

$$B_{\text{transition}} = D\sqrt{\frac{C_p}{C}} \qquad 40.34$$

Figure 40.8 *Pipes in Backfilled Trenches*

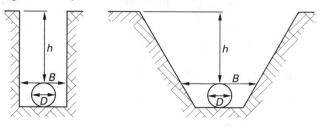

12. ALLOWABLE PIPE LOADS

$$\text{crushing strength} = (\text{D-load strength})D_i \qquad 40.35$$

$$w_{\text{allowable}} = \left(\begin{array}{c} \text{known pipe} \\ \text{crushing strength} \end{array}\right)\left(\frac{\text{LF}}{F}\right) \quad [\text{analysis}] \qquad 40.36$$

18. LIQUEFACTION

$$\frac{\tau_{h,\text{ave}}}{\sigma_o'} \approx 0.65\left(\frac{a_{\max}}{g}\right)\left(\frac{\sigma_o}{\sigma_o'}\right)r_d \qquad 40.37$$

Structural

CERM Chapter 41
Determinate Statics

> Chapter, section, equation, figure, and table numbers correspond to CERM. For additional study material, go to the corresponding chapter and section number in CERM.

4. CONCENTRATED FORCES

Figure 41.1 *Components and Direction Angles of a Force*

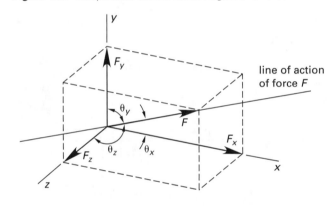

$$F_x = F \cos \theta_x \qquad 41.3$$

$$F_y = F \cos \theta_y \qquad 41.4$$

$$F_z = F \cos \theta_z \qquad 41.5$$

$$F = \sqrt{F_x^2 + F_y^2 + F_z^2} \qquad 41.6$$

6. MOMENT OF A FORCE ABOUT A POINT

$$\mathbf{M_O} = \mathbf{r} \times \mathbf{F} \qquad 41.7$$

$$M_O = |\mathbf{M_O}| = |\mathbf{r}|\,|\mathbf{F}| \sin \theta = d|\mathbf{F}| \quad [\theta \le 180°] \qquad 41.8$$

9. COMPONENTS OF A MOMENT

$$M_x = M \cos \theta_x \qquad 41.12$$

$$M_y = M \cos \theta_y \qquad 41.13$$

$$M_z = M \cos \theta_z \qquad 41.14$$

$$M_x = yF_z - zF_y \qquad 41.15$$

$$M_y = zF_x - xF_z \qquad 41.16$$

$$M_z = xF_y - yF_x \qquad 41.17$$

$$M = \sqrt{M_x^2 + M_y^2 + M_z^2} \qquad 41.18$$

10. COUPLES

Figure 41.4 *Couple*

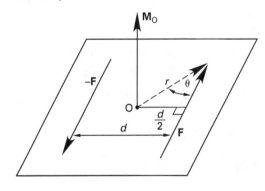

$$M_O = 2rF \sin \theta = Fd \qquad 41.19$$

15. MOMENT FROM A DISTRIBUTED LOAD

$$M_{\text{distributed load}} = \text{force} \times \text{distance}$$
$$= wx\left(\frac{x}{2}\right) = \tfrac{1}{2}\,wx^2 \qquad 41.34$$

17. CONDITIONS OF EQUILIBRIUM

$$\mathbf{F}_R = \sum \mathbf{F} = 0 \qquad 41.35$$

$$F_R = \sqrt{F_{R,x}^2 + F_{R,y}^2 + F_{R,z}^2} = 0 \qquad 41.36$$

$$\mathbf{M}_R = \sum \mathbf{M} = 0 \qquad 41.37$$

$$M_R = \sqrt{M_{R,x}^2 + M_{R,y}^2 + M_{R,z}^2} = 0 \qquad 41.38$$

25. INFLUENCE LINES FOR REACTIONS

$$R = F \times \text{influence line ordinate} \qquad 41.45$$

27. LEVERS

$$\frac{\text{mechanical}}{\text{advantage}} = \frac{F_{\text{load}}}{F_{\text{applied}}} = \frac{\text{applied force lever arm}}{\text{load lever arm}}$$
$$= \frac{\text{distance moved by applied force}}{\text{distance moved by load}} \qquad 41.46$$

32. DETERMINATE TRUSSES

$$\text{no. of members} = 2(\text{no. of joints}) - 3 \qquad 41.48$$

no. of members
+no. of reactions
$$- 2(\text{no. of joints}) = 0 \quad \text{[determinate]}$$
$$> 0 \quad \text{[indeterminate]}$$
$$< 0 \quad \text{[unstable]} \qquad 41.49$$

40. PARABOLIC CABLES

Figure 41.18 *Parabolic Cable*

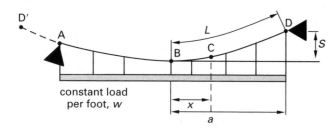

$$\sum M_{\text{D}} = wa\left(\frac{a}{2}\right) - HS = 0 \qquad 41.50$$

$$H = \frac{wa^2}{2S} \qquad 41.51$$

$$T_{\text{C},x} = H = \frac{wa^2}{2S} \qquad 41.52$$

$$T_{\text{C},y} = wx \qquad 41.53$$

$$T_{\text{C}} = \sqrt{(T_{\text{C},x})^2 + (T_{\text{C},y})^2}$$
$$= w\sqrt{\left(\frac{a^2}{2S}\right)^2 + x^2} \qquad 41.54$$

$$\tan\theta = \frac{wx}{H} \qquad 41.55$$

$$y(x) = \frac{wx^2}{2H} \qquad 41.56$$

$$L \approx a\left(1 + \left(\frac{2}{3}\right)\left(\frac{S}{a}\right)^2 - \left(\frac{2}{5}\right)\left(\frac{S}{a}\right)^4\right) \qquad 41.57$$

42. CATENARY CABLES

$$y(x) = c\,\cosh\left(\frac{x}{c}\right) \qquad 41.64$$

Figure 41.20 *Catenary Cable*

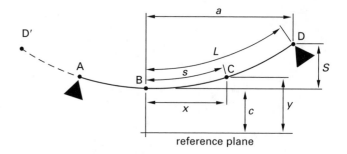

$$y = \sqrt{s^2 + c^2} = c\left(\cosh\left(\frac{x}{c}\right)\right) \qquad 41.65$$

$$s = c\left(\sinh\left(\frac{x}{c}\right)\right) \qquad 41.66$$

$$\text{sag} = S = y_{\text{D}} - c = c\left(\cosh\left(\frac{a}{c}\right) - 1\right) \qquad 41.67$$

$$\tan\theta = \frac{s}{c} \qquad 41.68$$

$$H = wc \qquad 41.69$$

$$F = ws \qquad 41.70$$

$$T = wy \qquad 41.71$$

$$\tan\theta = \frac{ws}{H} \qquad 41.72$$

$$\cos\theta = \frac{H}{T} \qquad 41.73$$

CERM Chapter 42
Properties of Areas

Chapter, section, equation, figure, and table numbers correspond to CERM. For additional study material, go to the corresponding chapter and section number in CERM.

1. CENTROID OF AN AREA

$$x_c = \frac{\int x \, dA}{A} \qquad 42.1$$

$$y_c = \frac{\int y \, dA}{A} \qquad 42.2$$

$$A = \int f(x) \, dx \qquad 42.3$$

$$x_c = \frac{\sum_i A_i x_{ci}}{\sum_i A_i} \qquad 42.5$$

$$y_c = \frac{\sum_i A_i y_{ci}}{\sum_i A_i} \qquad 42.6$$

2. FIRST MOMENT OF THE AREA

$$Q_y = \int x \, dA = x_c A \qquad 42.7$$

$$Q_x = \int y \, dA = y_c A \qquad 42.8$$

3. CENTROID OF A LINE

$$x_c = \frac{\int x \, dL}{L} \qquad 42.9$$

$$y_c = \frac{\int y \, dL}{L} \qquad 42.10$$

$$dL = \left(\sqrt{\left(\frac{dy}{dx}\right)^2 + 1} \right) dx \qquad 42.11$$

$$dL = \left(\sqrt{\left(\frac{dx}{dy}\right)^2 + 1} \right) dy \qquad 42.12$$

5. MOMENT OF INERTIA OF AN AREA

$$I_x = \int y^2 \, dA \qquad 42.17$$

$$I_y = \int x^2 \, dA \qquad 42.18$$

6. PARALLEL AXIS THEOREM

$$I_{\text{parallel axis}} = I_c + A d^2 \qquad 42.20$$

7. POLAR MOMENT OF INERTIA

$$J = \int (x^2 + y^2) \, dA \qquad 42.21$$

$$J = I_x + I_y \qquad 42.22$$

$$J = I_{cx} + I_{cy} \qquad 42.23$$

8. RADIUS OF GYRATION

$$I = r^2 A \qquad 42.24$$

$$r = \sqrt{\frac{I}{A}} \qquad 42.25$$

$$r = \sqrt{\frac{J}{A}} \qquad 42.26$$

$$r^2 = r_x^2 + r_y^2 \qquad 42.27$$

10. SECTION MODULUS

$$S = \frac{I_c}{c} \qquad 42.30$$

Centroids and Area Moments of Inertia for Basic Shapes

shape		centroidal location x_c	centroidal location y_c	area, A	area moment of inertia (rectangular and polar), I, J	radius of gyration, r
rectangle		$\dfrac{b}{2}$	$\dfrac{h}{2}$	bh	$I_x = \dfrac{bh^3}{3}$ $I_{c,x} = \dfrac{bh^3}{12}$ $J_c = \left(\dfrac{1}{12}\right)bh(b^2 + h^2)*$	$r_x = \dfrac{h}{\sqrt{3}}$ $r_{c,x} = \dfrac{h}{2\sqrt{3}}$
triangular area			$\dfrac{h}{3}$	$\dfrac{bh}{2}$	$I_x = \dfrac{bh^3}{12}$ $I_{c,x} = \dfrac{bh^3}{36}$	$r_x = \dfrac{h}{\sqrt{6}}$ $r_{c,x} = \dfrac{h}{3\sqrt{2}}$
trapezoid			$h\left(\dfrac{b+2t}{3b+3t}\right)$	$\dfrac{(b+t)h}{2}$	$I_x = \dfrac{(b+3t)h^3}{12}$ $I_{c,x} = \dfrac{(b^2+4bt+t^2)h^3}{36(b+t)}$	$r_x = \left(\dfrac{h}{\sqrt{6}}\right)\sqrt{\dfrac{b+3t}{b+t}}$ $r_{c,x} = \dfrac{h\sqrt{2(b^2+4bt+t^2)}}{6(b+t)}$
circle		0	0	πr^2	$I_x = I_y = \dfrac{\pi r^4}{4}$ $J_c = \dfrac{\pi r^4}{2}$	$r_x = \dfrac{r}{2}$
quarter-circular area		$\dfrac{4r}{3\pi}$	$\dfrac{4r}{3\pi}$	$\dfrac{\pi r^2}{4}$	$I_x = I_y = \dfrac{\pi r^4}{16}$ $J_o = \dfrac{\pi r^4}{8}$	
semicircular area		0	$\dfrac{4r}{3\pi}$	$\dfrac{\pi r^2}{2}$	$I_x = I_y = \dfrac{\pi r^4}{8}$ $I_{c,x} = 0.1098r^4$ $J_o = \dfrac{\pi r^4}{4}$ $J_c = 0.5025r^4$	$r_x = \dfrac{r}{2}$ $r_{c,x} = 0.264r$
quarter-elliptical area		$\dfrac{4a}{3\pi}$	$\dfrac{4b}{3\pi}$	$\dfrac{\pi ab}{4}$	$I_x = \dfrac{\pi ab^3}{8}$ $I_y = \dfrac{\pi a^3 b}{8}$	
semielliptical area		0	$\dfrac{4b}{3\pi}$	$\dfrac{\pi ab}{2}$	$J_o = \dfrac{\pi ab(a^2+b^2)}{8}$	
semiparabolic area		$\dfrac{3a}{8}$	$\dfrac{3h}{5}$	$\dfrac{2ah}{3}$		
parabolic area		0	$\dfrac{3h}{5}$	$\dfrac{4ah}{3}$	$I_x = \dfrac{4ah^3}{7}$ $I_y = \dfrac{4ha^3}{15}$ $I_{c,x} = \dfrac{16ah^3}{175}$	$r_x = h\sqrt{\dfrac{3}{7}}$ $r_y = \dfrac{a}{\sqrt{5}}$
parabolic spandrel		$\dfrac{3a}{4}$	$\dfrac{3h}{10}$	$\dfrac{ah}{3}$	$I_x = \dfrac{ah^3}{21}$ $I_y = \dfrac{3ha^3}{15}$	
general spandrel		$\left(\dfrac{n+1}{n+2}\right)a$	$\left(\dfrac{n+1}{4n+2}\right)h$	$\dfrac{ah}{n+1}$	*Theoretical definition based on $J = I_x + I_y$. However, in torsion, not all parts of the shape are effective. Effective values will be lower.	
circular sector [α in radians]		$\dfrac{2r\sin\alpha}{3\alpha}$	0	αr^2	$J = C\left(\dfrac{b^2+h^2}{b^3 h^3}\right)$	

b/h	C
1	3.56
2	3.50
4	3.34
8	3.21

CERM Chapter 43
Material Properties and Testing

Chapter, section, equation, figure, and table numbers correspond to CERM. For additional study material, go to the corresponding chapter and section number in CERM.

1. TENSILE TEST

$$s = \frac{F}{A_o} \qquad 43.1$$

$$e = \frac{\delta}{L_o} \qquad 43.2$$

$$s = Ee \qquad 43.3$$

5. POISSON'S RATIO

$$\nu = \frac{e_{\text{lateral}}}{e_{\text{axial}}} = \frac{\dfrac{\Delta D}{D_o}}{\dfrac{\delta}{L_o}} \qquad 43.4$$

7. TRUE STRESS AND STRAIN

$$\sigma = \frac{F}{A} = \frac{F}{\left(1 - \dfrac{A_o - A}{A_o}\right)A_o} = \frac{F}{(1 - q)\,A_o}$$

$$= \frac{s}{(1 - e\nu)^2} = s(1 + e) \qquad 43.5$$

$$\epsilon = \int_{L_0}^{L} \frac{dL}{L} = \ln\left(\frac{L}{L_o}\right)$$

$$= \ln(1 + e) \quad \text{[prior to necking]} \qquad 43.6$$

$$A_o L_o = AL \qquad 43.7$$

$$\epsilon = \ln\left(\frac{A_o}{A}\right) = \ln\left(\frac{D_o}{D}\right)^2 = 2\ln\left(\frac{D_o}{D}\right) \qquad 43.8$$

$$\sigma = K\epsilon^n \qquad 43.9$$

8. DUCTILITY

$$\text{percent elongation} = \frac{L_f - L_o}{L_o} \times 100\%$$

$$= e_f \times 100\% \qquad 43.10$$

$$\text{ductility} = \frac{\text{ultimate failure strain}}{\text{yielding strain}} \qquad 43.11$$

$$\text{reduction in area} = \frac{A_o - A_f}{A_o} \times 100\% \qquad 43.12$$

9. STRAIN ENERGY

$$\text{work} = \text{force} \times \text{distance} = \int F\,dL \qquad 43.13$$

$$\frac{\text{work per}}{\text{unit volume}} = \int \frac{F\,dL}{AL} = \int_0^{\epsilon_{\text{final}}} \sigma\,d\epsilon \qquad 43.14$$

10. RESILIENCE

$$U_R = \int_0^{\epsilon_y} \sigma\,d\epsilon = E\int_0^{\epsilon_y} \epsilon\,d\epsilon = \frac{E\epsilon_y^2}{2} = \frac{S_y\epsilon_y}{2} \qquad 43.15$$

11. TOUGHNESS

$$U_T \approx S_u\epsilon_u \quad \text{[ductile]} \qquad 43.16$$

$$U_T \approx \left(\frac{S_y + S_u}{2}\right)\epsilon_u \quad \text{[ductile]} \qquad 43.17$$

$$U_T \approx \tfrac{2}{3}S_u\epsilon_u \quad \text{[brittle]} \qquad 43.18$$

14. TORSION TEST

$$\tau = G\theta \qquad 43.20$$

$$G = \frac{E}{2(1 + \nu)} \qquad 43.21$$

$$\gamma = \frac{TL}{JG} = \frac{\tau L}{rG} \quad \text{[radians]} \qquad 43.22$$

$$S_{ys} = \frac{S_{yt}}{\sqrt{3}} = 0.577 S_{yt} \qquad 43.23$$

CERM Chapter 44
Strength of Materials

Chapter, section, equation, figure, and table numbers correspond to CERM. For additional study material, go to the corresponding chapter and section number in CERM.

2. HOOKE'S LAW

$$\sigma = E\epsilon \qquad 44.2$$

$$\tau = G\phi \qquad 44.3$$

3. ELASTIC DEFORMATION

$$\delta = L_o \epsilon = \frac{L_o \sigma}{E} = \frac{L_o F}{EA} \qquad 44.4$$

$$L = L_o + \delta \qquad 44.5$$

4. TOTAL STRAIN ENERGY

$$U = \tfrac{1}{2} F \delta = \frac{F^2 L_o}{2AE} = \frac{\sigma^2 L_o A}{2E} \qquad 44.6$$

5. STIFFNESS AND RIGIDITY

$$k = \frac{F}{\delta} \quad \text{[general form]} \qquad 44.7(a)$$

$$k = \frac{AE}{L_o} \quad \text{[normal stress form]} \qquad 44.7(b)$$

$$R_j = \frac{k_j}{\displaystyle\sum_i k_i} \qquad 44.8$$

6. THERMAL DEFORMATION

$$\Delta L = \alpha L_o (T_2 - T_1) \qquad 44.9$$

$$\Delta A = \gamma A_o (T_2 - T_1) \qquad 44.10$$

$$\gamma \approx 2\alpha \qquad 44.11$$

$$\Delta V = \beta V_o (T_2 - T_1) \qquad 44.12$$

$$\beta \approx 3\alpha \qquad 44.13$$

$$\epsilon_{\text{th}} = \frac{\Delta L}{L_o} = \alpha (T_2 - T_1) \qquad 44.14$$

$$\sigma_{\text{th}} = E \epsilon_{\text{th}} \qquad 44.15$$

7. STRESS CONCENTRATIONS

$$\sigma' = K \sigma_0 \qquad 44.16$$

13. SHEAR STRESS IN BEAMS

$$\tau = \frac{V}{A} \qquad 44.28$$

$$\tau = \frac{V}{t_w d} \qquad 44.29$$

Figure 44.10 *Web of a Flanged Beam*

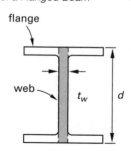

Figure 44.11 *Shear Stress Distribution Within a Rectangular Beam*

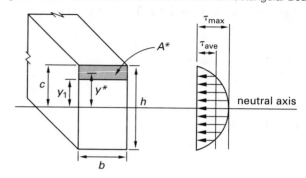

$$Q = y^* A^* \qquad 44.32$$

$$\tau_{\text{max,rectangular}} = \frac{3V}{2A} = \frac{3V}{2bh} \qquad 44.33$$

$$\tau_{\text{max,circular}} = \frac{4V}{3A} = \frac{4V}{3\pi r^2} \qquad 44.34$$

$$\tau_{\text{max,hollow cylinder}} = \frac{2V}{A} \qquad 44.35$$

14. BENDING STRESS IN BEAMS

$$\sigma_b = \frac{-My}{I_c} \qquad 44.36$$

Figure 44.13 *Bending Stress Distribution in a Beam*

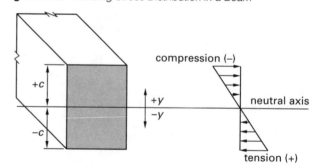

$$\sigma_{b,\text{max}} = \frac{Mc}{I_c} \qquad 44.37$$

$$\sigma_{b,\text{max}} = \frac{M}{S} \qquad 44.38$$

$$S = \frac{I_c}{c} \qquad 44.39$$

$$S_{\text{rectangular}} = \frac{bh^2}{6} \qquad 44.40$$

15. STRAIN ENERGY DUE TO BENDING MOMENT

$$U = \frac{1}{2EI} \int M^2(x)\, dx \qquad 44.41$$

16. ECCENTRIC LOADING OF AXIAL MEMBERS

Figure 44.14 Eccentric Loading of an Axial Member

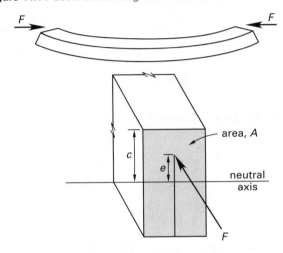

$$\sigma_{\text{max,min}} = \frac{F}{A} \pm \frac{Mc}{I_c} \qquad 44.42$$

$$= \frac{F}{A} \pm \frac{Fec}{I_c} \qquad 44.43$$

17. BEAM DEFLECTION: DOUBLE INTEGRATION METHOD

$$y = \text{deflection} \qquad 44.44$$

$$y' = \frac{dy}{dx} = \text{slope} \qquad 44.45$$

$$y'' = \frac{d^2y}{dx^2} = \frac{M(x)}{EI} \qquad 44.46$$

$$y''' = \frac{d^3y}{dx^3} = \frac{V(x)}{EI} \qquad 44.47$$

$$y = \frac{1}{EI} \int \left(\int M(x)\, dx \right) dx \qquad 44.48$$

18. BEAM DEFLECTION: MOMENT AREA METHOD

$$\phi = \int \frac{M(x)\, dx}{EI} \qquad 44.49$$

$$y = \int \frac{x M(x)\, dx}{EI} \qquad 44.50$$

19. BEAM DEFLECTION: STRAIN ENERGY METHOD

$$\tfrac{1}{2} Fy = \sum U \qquad 44.51$$

23. INFLECTION POINTS

$$y''(x) = \frac{1}{\rho(x)} = \frac{M(x)}{EI} \qquad 44.52$$

25. TRUSS DEFLECTION: VIRTUAL WORK METHOD

$$\delta = \sum \frac{SuL}{AE} \qquad 44.53$$

28. COMPOSITE STRUCTURES

$$n = \frac{E}{E_{\text{weakest}}} \qquad 44.54$$

$$\sigma_{\text{weakest}} = \frac{F}{A_t} \qquad 44.55$$

$$\sigma_{\text{stronger}} = \frac{nF}{A_t} \qquad 44.56$$

$$\sigma_{\text{weakest}} = \frac{Mc_{\text{weakest}}}{I_{c,t}} \qquad 44.57$$

$$\sigma_{\text{stronger}} = \frac{nMc_{\text{stronger}}}{I_{c,t}} \qquad 44.58$$

CERM Chapter 45
Basic Elements of Design

> Chapter, section, equation, figure, and table numbers correspond to CERM. For additional study material, go to the corresponding chapter and section number in CERM.

1. SLENDER COLUMNS

r is the radius of gyration.

$$F_e = \frac{\pi^2 EI}{L^2} = \frac{\pi^2 EA}{\left(\frac{L}{r}\right)^2} \qquad 45.1$$

$$\sigma_e = \frac{F_e}{A} = \frac{\pi^2 E}{\left(\frac{L}{r}\right)^2} \qquad 45.2$$

$$L' = KL \qquad 45.3$$

$$\sigma_e = \frac{F_e}{A} = \frac{\pi^2 E}{\left(\frac{KL}{r}\right)^2} \qquad 45.4$$

Table 45.1 Theoretical End Restraint Coefficients

illus.	end conditions	K ideal	recommended for design
(a)	both ends pinned	1	1.0*
(b)	both ends built in	0.5	0.65*–0.90
(c)	one end pinned, one end built in	0.707	0.80*–0.90
(d)	one end built in, one end free	2	2.0–2.1*
(e)	one end built in, one end fixed against rotation but free	1	1.2*
(f)	one end pinned, one end fixed against rotation but free	2	2.0*

* AISC values

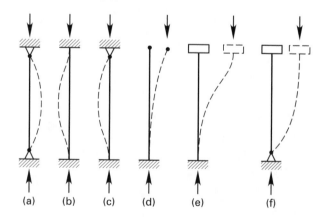

(a) (b) (c) (d) (e) (f)

Figure 45.1 Euler's Curve

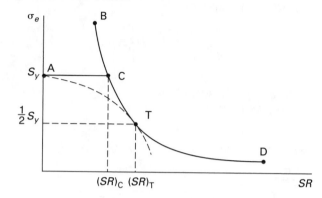

$$(SR)_T = \sqrt{\frac{2\pi^2 KE}{S_y}} \qquad 45.5$$

2. INTERMEDIATE COLUMNS

$$\sigma_{cr} = \frac{P_{cr}}{A} = a - b\left(\frac{KL}{r}\right)^2 \qquad 45.6$$

$$\sigma_{cr} = S_y - \left(\frac{1}{E}\right)\left(\frac{S_y}{2\pi}\right)^2\left(\frac{KL}{r}\right)^2 \qquad 45.7$$

3. ECCENTRICALLY LOADED COLUMNS

$$\sigma_{max} = \sigma_{ave}(1 + \text{amplification factor})$$

$$= \left(\frac{F}{A}\right)\left(1 + \left(\frac{ec}{r^2}\right)\sec\left(\frac{\pi}{2}\sqrt{\frac{F}{F_e}}\right)\right)$$

$$= \left(\frac{F}{A}\right)\left(1 + \left(\frac{ec}{r^2}\right)\sec\left(\frac{L}{2r}\sqrt{\frac{F}{AE}}\right)\right)$$

$$= \left(\frac{F}{A}\right)\left(1 + \left(\frac{ec}{r^2}\right)\sec\phi\right) \qquad 45.8$$

$$\phi = \tfrac{1}{2}\left(\frac{L}{r}\right)\sqrt{\frac{F}{AE}} \qquad 45.9$$

4. THIN-WALLED CYLINDRICAL TANKS

$$\frac{t}{d_i} = \frac{t}{2r_i} < 0.1 \qquad \text{[thin-walled]} \qquad 45.10$$

Figure 45.2 Stresses in a Thin-Walled Tank

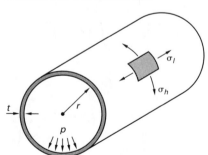

$$\sigma_h = \frac{pr}{t} \qquad \text{45.11}$$

$$\sigma_l = \frac{pr}{2t} \qquad \text{45.12}$$

$$\Delta L = L\epsilon_l$$
$$= L\left(\frac{\sigma_l - \nu\sigma_h}{E}\right) \qquad \text{45.13}$$

$$\Delta C = C\epsilon_h$$
$$= \pi d_o\left(\frac{\sigma_h - \nu\sigma_l}{E}\right) \qquad \text{45.14}$$

5. THICK-WALLED CYLINDERS

$$\sigma_c = \frac{r_i^2 p_i - r_o^2 p_o + \dfrac{(p_i - p_o)\, r_i^2 r_o^2}{r^2}}{r_o^2 - r_i^2} \qquad \text{45.15}$$

Figure 45.3 Thick-Walled Cylinder

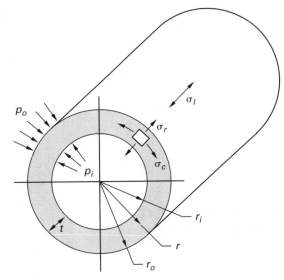

$$\sigma_r = \frac{r_i^2 p_i - r_o^2 p_o - \dfrac{(p_i - p_o)\, r_i^2 r_o^2}{r^2}}{r_o^2 - r_i^2} \qquad \text{45.16}$$

$$\sigma_l = \frac{p_i r_i^2}{r_o^2 - r_i^2} \qquad \left[\begin{array}{c} p_o \text{ does not act} \\ \text{longitudinally on the ends} \end{array}\right] \text{45.17}$$

Table 45.2 Stresses in Thick-Walled Cylinders[a]

stress	external pressure, p	internal pressure, p
$\sigma_{c,o}$	$\dfrac{-(r_o^2 + r_i^2)\,p}{r_o^2 - r_i^2}$	$\dfrac{2r_i^2 p}{r_o^2 - r_i^2}$
$\sigma_{r,o}$	$-p$	0
$\sigma_{c,i}$	$\dfrac{-2r_o^2 p}{r_o^2 - r_i^2}$	$\dfrac{(r_o^2 + r_i^2)\,p}{r_o^2 - r_i^2}$
$\sigma_{r,i}$	0	$-p$
$\tau_{\max}$	$\frac{1}{2}\sigma_{c,i}$	$\frac{1}{2}(\sigma_{c,i} + p)$

[a]Table 45.2 can be used with thin-walled cylinders. However, in most cases it will not be necessary to do so.

$$\epsilon = \frac{\Delta d}{d} = \frac{\Delta C}{C} = \frac{\Delta r}{r}$$
$$= \frac{\sigma_c - \nu(\sigma_r + \sigma_l)}{E} \qquad \text{45.18}$$

6. THIN-WALLED SPHERICAL TANKS

$$\sigma = \frac{pr}{2t} \qquad \text{45.19}$$

7. INTERFERENCE FITS

$$I_{\text{diametral}} = 2I_{\text{radial}}$$
$$= d_{o,\text{inner}} - d_{i,\text{outer}}$$
$$= |\Delta d_{o,\text{inner}}| + |\Delta d_{i,\text{outer}}| \qquad \text{45.20}$$

$$\epsilon = \frac{\Delta d}{d} = \frac{\Delta C}{C} = \frac{\Delta r}{r}$$
$$= \frac{\sigma_c - \nu\sigma_r}{E} \qquad \text{45.21}$$

$$I_{\text{diametral}} = 2I_{\text{radial}}$$
$$= \left(\frac{2pr_{o,\text{shaft}}}{E_{\text{hub}}}\right)\left(\frac{r_{o,\text{hub}}^2 + r_{o,\text{shaft}}^2}{r_{o,\text{hub}}^2 - r_{o,\text{shaft}}^2} + \nu_{\text{hub}}\right)$$
$$+ \left(\frac{2pr_{o,\text{shaft}}}{E_{\text{shaft}}}\right)\left(\frac{r_{o,\text{shaft}}^2 + r_{i,\text{shaft}}^2}{r_{o,\text{shaft}}^2 - r_{i,\text{shaft}}^2} - \nu_{\text{shaft}}\right)$$
$$\text{45.22}$$

$$I_{\text{diametral}} = 2I_{\text{radial}}$$
$$= \left(\frac{4pr_{\text{shaft}}}{E}\right)\left(\frac{1}{1 - \left(\dfrac{r_{\text{shaft}}}{r_{o,\text{hub}}}\right)^2}\right) \qquad \text{45.23}$$

$$F_{max} = fN = 2\pi f p r_{o,\text{shaft}} L_{\text{interface}} \qquad 45.24$$

$$T_{max} = 2\pi f p r_{o,\text{shaft}}^2 L_{\text{interface}} \qquad 45.25$$

10. RIVET AND BOLT CONNECTIONS

$$\tau = \frac{F}{\frac{\pi}{4}d^2} \qquad 45.26$$

$$A_t = t(b - nd) \qquad 45.28$$

$$\sigma_t = \frac{F}{A_t} \qquad 45.29$$

$$\sigma_p = \frac{F}{dt} \qquad 45.30$$

$$n = \frac{\sigma_p}{\text{allowable bearing stress}} \qquad 45.31$$

$$\tau = \frac{F}{2A} = \frac{F}{2t\left(L - \dfrac{d}{2}\right)} \qquad 45.32$$

11. BOLT PRELOAD

$$k_{\text{bolt}} = \frac{F}{\Delta L} = \frac{A_{\text{bolt}} E_{\text{bolt}}}{L_{\text{bolt}}} \qquad 45.33$$

$$k_{\text{parts}} = \frac{A_{e,\text{parts}} E_{\text{parts}}}{L_{\text{parts}}} \qquad 45.34$$

$$\frac{1}{k_{\text{parts,composite}}} = \frac{1}{k_1} + \frac{1}{k_2} + \frac{1}{k_3} + \cdots \qquad 45.35$$

$$F_{\text{bolt}} = F_i + \frac{k_{\text{bolt}} F_{\text{applied}}}{k_{\text{bolt}} + k_{\text{parts}}} \qquad 45.36$$

$$F_{\text{parts}} = \frac{k_{\text{parts}} F_{\text{applied}}}{k_{\text{bolt}} + k_{\text{parts}}} - F_i \qquad 45.37$$

$$\sigma_{\text{bolt}} = \frac{KF}{A} \qquad 45.38$$

12. BOLT TORQUE TO OBTAIN PRELOAD

$$T = K_T d_{\text{bolt}} F_i \qquad 45.39$$

$$K_T = \frac{f_c r_c}{d_{\text{bolt}}} + \left(\frac{r_t}{d_{\text{bolt}}}\right)\left(\frac{\tan\theta + f_t \sec\alpha}{1 - f_t \tan\theta \sec\alpha}\right) \qquad 45.40$$

$$\tan\theta = \frac{\text{lead per revolution}}{2\pi r_t} \qquad 45.41$$

13. FILLET WELDS

$$t_e = 0.707y \qquad 45.42$$

$$\tau = \frac{F}{bt_e} \qquad 45.43$$

14. CIRCULAR SHAFT DESIGN

$$\tau = G\theta = \frac{Tr}{J} \qquad 45.44$$

$$U = \frac{T^2 L}{2GJ} \qquad 45.45$$

$$J = \frac{\pi r^4}{2} = \frac{\pi d^4}{32} \qquad 45.46$$

$$J = \frac{\pi}{2}\left(r_o^4 - r_i^4\right) \qquad 45.47$$

$$\gamma = \frac{L\theta}{r} = \frac{TL}{GJ} \qquad 45.48$$

$$G = \frac{E}{2(1 + \nu)} \qquad 45.49$$

$$T_{\text{in-lbf}} = \frac{(63{,}025)(\text{horsepower})}{n_{\text{rpm}}} \qquad 45.50$$

$$\tau_{max} = \sqrt{\left(\frac{\sigma_x}{2}\right)^2 + \tau^2} \qquad 45.51$$

$$\tau_{max} = \frac{16}{\pi d^3}\sqrt{M^2 + T^2} \qquad 45.52$$

$$\sigma' = \frac{16}{\pi d^3}\sqrt{4M^2 + 3T^2} \qquad 45.53$$

15. TORSION IN THIN-WALLED, NONCIRCULAR SHELLS

$$\tau = \frac{T}{2At} \qquad 45.54$$

$$q = \tau t = \frac{T}{2A} \quad [\text{constant}] \qquad 45.55$$

$$\gamma = \frac{TLp}{4A^2 tG} \qquad 45.56$$

18. ECCENTRICALLY LOADED BOLTED CONNECTIONS

$$\tau = \frac{Tr}{J} = \frac{Fer}{J} \qquad 45.57$$

$$J = \sum_i r_i^2 A_i \qquad 45.58$$

$$\tau_v = \frac{F}{nA} \qquad 45.59$$

21. SPRINGS

$$F = k\delta \qquad 45.60$$

$$k = \frac{F_1 - F_2}{\delta_1 - \delta_2} \qquad 45.61$$

$$\Delta E_p = \tfrac{1}{2}k\delta^2 \qquad 45.62$$

$$m\left(\frac{g}{g_c}\right)(h + \delta) = \tfrac{1}{2}k\delta^2 \qquad 45.63(b)$$

$$\frac{1}{k_{\text{eq}}} = \frac{1}{k_1} + \frac{1}{k_2} + \frac{1}{k_3} + \cdots \qquad 45.64$$

$$k_{\text{eq}} = k_1 + k_2 + k_3 + \cdots \qquad 45.65$$

22. WIRE ROPE

$$\sigma_{\text{bending}} = \frac{d_w E_w}{d_{\text{sh}}} \qquad 45.66$$

$$p_{\text{bearing}} = \frac{2F_t}{d_r d_{\text{sh}}} \qquad 45.67$$

CERM Chapter 46
Structural Analysis I

Chapter, section, equation, figure, and table numbers correspond to CERM. For additional study material, go to the corresponding chapter and section number in CERM.

4. REVIEW OF ELASTIC DEFORMATION[1]

$$\delta = \frac{FL}{AE} \qquad 46.2$$

$$\delta = \alpha L_o \Delta T \qquad 46.3$$

7. THREE-MOMENT EQUATION

Figure 46.2 Portion of a Continuous Beam

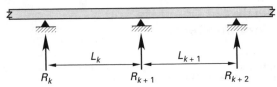

$$\frac{M_k L_k}{I_k} + (2M_{k+1})\left(\frac{L_k}{I_k} + \frac{L_{k+1}}{I_{k+1}}\right) + \frac{M_{k+2}L_{k+1}}{I_{k+1}}$$
$$= -6\left(\frac{A_k a}{I_k L_k} + \frac{A_{k+1} b}{I_{k+1} L_{k+1}}\right) \qquad 46.4$$

---

[1]This subject is covered in greater detail in CERM Ch. 44.

8. FIXED-END MOMENTS

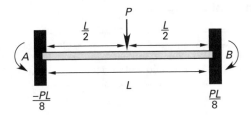

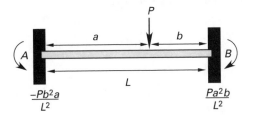

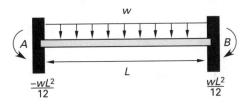

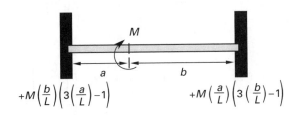

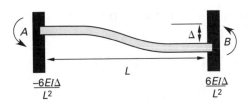

10. INFLUENCE DIAGRAMS

$$R_L = P \times \text{ordinate} \qquad 46.9$$

$$R_L = \int_{x_1}^{x_2} (w \times \text{ordinate})\,dx$$
$$= w \times \text{area under curve} \qquad 46.10$$

CERM Chapter 47
Structural Analysis II

> Chapter, section, equation, figure, and table numbers correspond to CERM. For additional study material, go to the corresponding chapter and section number in CERM.

3. REVIEW OF WORK AND ENERGY

$$W = P\Delta \quad \text{[linear displacement]} \qquad 47.1$$

$$W = T\phi \quad \text{[rotation]} \qquad 47.2$$

$$W = U_2 - U_1 \qquad 47.3$$

4. REVIEW OF LINEAR DEFORMATION

$$\Delta = \frac{PL}{AE} \qquad 47.4$$

5. THERMAL LOADING

$$P_{th} = \frac{\Delta_{\text{constrained}} AE}{L} = \alpha(T_2 - T_1)\left(\frac{LAE}{L}\right)$$
$$= \alpha(T_2 - T_1)AE \qquad 47.5$$

$$M = \alpha(T_{\text{extreme fiber}} - T_{\text{neutral axis}})\left(\frac{EI}{\frac{h}{2}}\right) \qquad 47.6$$

6. DUMMY UNIT LOAD METHOD

$$W_Q = 1 \times \Delta = \int \sigma_Q \epsilon_P dV \qquad 47.7$$

$$W_Q = W_m + W_v + W_a + W_t \qquad 47.8$$

$$W_m = \int \left(\frac{m_Q m_P}{EI}\right) ds \qquad 47.9$$

$$W_v = \int \left(\frac{V_Q V_P}{GA}\right) ds \qquad 47.10$$

$$W_a = \int \left(\frac{N_Q N_P}{EA}\right) ds \qquad 47.11$$

$$W_t = \int \left(\frac{T_Q T_P}{GJ}\right) ds \qquad 47.12$$

7. BEAM DEFLECTIONS BY THE DUMMY UNIT LOAD METHOD

$$W_Q = W_m = \int \left(\frac{m_Q m_P}{EI}\right) ds \qquad 47.13$$

23. APPROXIMATE METHOD: MOMENT COEFFICIENTS

$$M = C_1 w L^2 \qquad 47.43$$

Table 47.2 ACI Moment Coefficients

condition	C_1
positive moments near midspan	
end spans	
simply supported	$\frac{1}{11}$
built-in support	$\frac{1}{14}$
interior spans	$\frac{1}{16}$
negative moments at exterior face of first interior support	
2 spans	$\frac{1}{9}$
3 or more spans	$\frac{1}{10}$
negative moments at other faces or interior supports	
all cases	$\frac{1}{11}$
negative moments at face of all supports	
slabs with spans not exceeding 10 ft, and beams with ratio of sum of column stiffnesses to beam stiffness exceed 8 at each end of the span	$\frac{1}{12}$
negative moments at exterior built-in support	
support is a cross beam or girder (spandrel beam)	$\frac{1}{24}$
support is a column	$\frac{1}{16}$

24. APPROXIMATE METHOD: SHEAR COEFFICIENTS

$$V = C_2 \left(\frac{wL}{2}\right) \qquad 47.44$$

C_2 for shear in end members at the first interior support is 1.15. For shear at the face of all other supports, $C_2 = 1.0$.

CERM Chapter 48
Properties of Concrete and Reinforcing Steel

> Chapter, section, equation, figure, and table numbers correspond to CERM. For additional study material, go to the corresponding chapter and section number in CERM.

8. COMPRESSIVE STRENGTH

$$f'_c = \frac{P}{A} \qquad 48.1$$

10. MODULUS OF ELASTICITY

$$E_c = w_c^{1.5} 33\sqrt{f_c'} \qquad \left[90\,\frac{\text{lbf}}{\text{ft}^3} = w_c^{1.5} 33\sqrt{f_c'}\right] \quad 48.2(b)$$

$$E_c = 57{,}000\sqrt{f_c'} \quad \text{[normal-weight concrete]} \quad 48.3(b)$$

11. SPLITTING TENSILE STRENGTH

$$f_{\text{ct}} = \frac{2P}{\pi DL} \qquad\qquad 48.4$$

$$f_{\text{ct}} = 6.7\sqrt{f_c'} \quad \text{[normal weight]} \qquad 48.5$$

$$f_{\text{ct}} = 5.7\sqrt{f_c'} \quad \text{[sand lightweight]} \qquad 48.6$$

$$f_{\text{ct}} = 5\sqrt{f_c'} \quad \text{[all-lightweight]} \qquad 48.7$$

12. MODULUS OF RUPTURE

$$f_r = \frac{Mc}{I} \quad \text{[tension]} \qquad\qquad 48.8$$

$$f_r = 7.5\sqrt{f_c'} \qquad\qquad 48.9(b)$$

Table 49.2 *Summary of Approximate Properties of Concrete Components*

cement	
specific weight	195 lbf/ft^3 (3120 kg/m^3)
specific gravity	3.13–3.15
weight of one sack	94 lbf (42 kg)
fine aggregate	
specific weight	165 lbf/ft^3 (2640 kg/m^3)
specific gravity	2.64
coarse aggregate	
specific weight	165 lbf/ft^3 (2640 kg/m^3)
specific gravity	2.64
water	
specific weight	62.4 lbf/ft^3 (1000 kg/m^3)
	7.48 gal/ft^3 (1000 L/m^3)
	8.34 lbf/gal (1 kg/L)
	239.7 gal/ton (1 L/kg)
specific gravity	1.00

(Multiply lbf/ft^3 by 16 to obtain kg/m^3.)
(Multiply lbf by 0.45 to obtain kg.)

CERM Chapter 49
Concrete Proportioning, Mixing, and Placing

> Chapter, section, equation, figure, and table numbers correspond to CERM. For additional study material, go to the corresponding chapter and section number in CERM.

6. ABSOLUTE VOLUME METHOD

$$V_{\text{absolute}} = \frac{W}{(\text{SG})\gamma_{\text{water}}} \qquad\qquad 49.1(b)$$

Table 49.3 *Dry and Wet Basis Calculations*

	dry basis	wet basis
fraction moisture, f	$\dfrac{W_{\text{excess water}}}{W_{\text{SSD sand}}}$	$\dfrac{W_{\text{excess water}}}{W_{\text{SSD sand}} + W_{\text{excess water}}}$
weight of sand, $W_{\text{wet sand}}$	$W_{\text{SSD sand}} + W_{\text{excess water}}$ $(1+f)W_{\text{SSD sand}}$	$W_{\text{SSD sand}} + W_{\text{excess water}}$ $\left(\dfrac{1}{1-f}\right)W_{\text{SSD sand}}$
weight of SSD sand, $W_{\text{SSD sand}}$	$\dfrac{W_{\text{wet sand}}}{1+f}$	$(1-f)W_{\text{wet sand}}$
weight of excess water, $W_{\text{excess water}}$	$fW_{\text{SSD sand}}$ $\dfrac{fW_{\text{wet sand}}}{1+f}$	$fW_{\text{wet sand}}$ $\dfrac{fW_{\text{SSD sand}}}{1-f}$

CERM Chapter 50
Reinforced Concrete: Beams

> Chapter, section, equation, figure, and table numbers correspond to CERM. For additional study material, go to the corresponding chapter and section number in CERM.

5. SERVICE LOADS, FACTORED LOADS, AND LOAD COMBINATIONS

$$U = 1.4D \qquad \text{50.1}$$

$$U = 1.2D + 1.6L \qquad \text{50.2}$$

$$U = 1.2D + 1.6L + 0.8W \qquad \text{50.3}$$

$$U = 1.2D + L + 1.6W \qquad \text{50.4}$$

$$U = 0.9D + 1.6W \qquad \text{50.5}$$

$$U = 1.2D + L + E \qquad \text{50.6}$$

$$U = 0.9D + E \qquad \text{50.7}$$

$$U = 1.4D + 1.7L \qquad \text{50.8}$$

$$U = 0.75(1.4D + 1.7L) + (1.6W \text{ or } E) \qquad \text{50.9}$$

$$U = 0.9D + (1.6W \text{ or } E) \qquad \text{50.10}$$

6. DESIGN STRENGTH AND DESIGN CRITERIA

$$\text{design strength} = (\text{nominal strength})\phi \qquad \text{50.11}$$

$$\phi M_n \geq M_u \qquad \text{50.12(a)}$$

$$\phi V_n \geq V_u \qquad \text{50.12(b)}$$

7. MINIMUM STEEL AREA

$$A_{s,\min} = \frac{3\sqrt{f_c'}b_w d}{f_y} \geq \frac{200 b_w d}{f_y} \qquad \text{50.16(b)}$$

$$A_{s,\min} = \frac{6\sqrt{f_c'}b_w d}{f_y} \qquad \text{50.17(b)}$$

8. MAXIMUM STEEL AREA

$$A_{s,\max} = 0.75 A_{sb} \qquad \text{50.18}$$

$$A_{sb} = \frac{0.85 f_c' A_{cb}}{f_y} \qquad \text{50.19}$$

$$a_b = \beta_1 \left(\frac{87,000}{87,000 + f_y} \right) d \qquad \text{50.20(b)}$$

$$\beta_1 = 0.85 \quad [f_c' \leq 4000 \text{ lbf/in}^2] \qquad \text{50.21(b)}$$

$$\beta_1 = 0.85 - (0.05)\left(\frac{f_c' - 4000}{1000} \right) \geq 0.65 \qquad \text{50.21(d)}$$

$$A_{sb} = \left(\left(\frac{0.85 \beta_1 f_c'}{f_y} \right) \left(\frac{87,000}{87,000 + f_y} \right) \right) bd \qquad \text{50.22(b)}$$

Figure 50.5 Beam at Balanced Condition

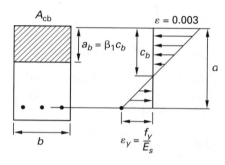

9. STEEL COVER AND BEAM WIDTH

$$1.5 \leq d/b \leq 2.5 \qquad \text{50.23}$$

Table 50.2 Minimum Beam Widths[a,b] for Beams with 1½ in Cover (inches)

size of bar	number of bars in a single layer of reinforcement							add for each additional bar[c]
	2	3	4	5	6	7	8	
no. 4	6.8	8.3	9.8	11.3	12.8	14.3	15.8	1.50
no. 5	6.9	8.5	10.2	11.8	13.4	15.0	16.7	1.63
no. 6	7.0	8.8	10.5	12.3	14.0	15.8	17.5	1.75
no. 7	7.2	9.0	10.9	12.8	14.7	16.5	18.4	1.88
no. 8	7.3	9.3	11.3	13.3	15.3	17.3	19.3	2.00
no. 9	7.6	9.8	12.2	14.3	16.6	18.8	21.1	2.26
no. 10	7.8	10.4	12.9	15.5	18.0	20.5	23.1	2.54
no. 11	8.1	10.9	13.8	16.6	19.4	22.2	25.0	2.82
no. 14	8.9	12.3	15.7	19.1	22.5	25.9	29.3	3.40
no. 18	10.6	15.1	19.6	24.1	28.6	33.1	37.6	4.51

(Multiply in by 25.4 to obtain mm.)
[a]Using no. 3 stirrups. If stirrups are not used, deduct 0.75 in.
[b]The minimum inside radius of a 90° stirrup bend is 2 times the stirrup diameter. An allowance has been included in the beam widths to achieve a full bend radius.
[c]For additional horizontal bars, the beam width is increased by adding the value in the last column.

10. NOMINAL MOMENT STRENGTH OF SINGLY REINFORCED SECTIONS

$$M_n = f_y A_s \times \text{lever arm} \qquad 50.24$$

$$T = f_y A_s \qquad 50.25$$

$$A_c = \frac{T}{0.85 f_c'} = \frac{f_y A_s}{0.85 f_c'} \qquad 50.26$$

$$M_n = A_s f_y (d - \lambda) \qquad 50.27$$

Figure 50.6 *Conditions at Maximum Moment*

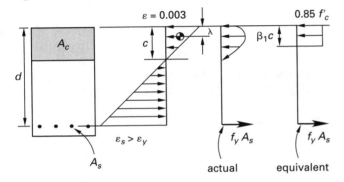

(a) strain distribution

(b) compressive stress distribution

(c) equivalent rectangular compressive stress block

11. BEAM DESIGN: SIZE KNOWN, REINFORCEMENT UNKNOWN

$$A_s = \frac{M_u}{\phi f_y (d - \lambda)} \qquad 50.28$$

12. BEAM DESIGN: SIZE AND REINFORCEMENT UNKNOWN

$$\lambda = \frac{A_c}{2b} \qquad 50.29$$

$$M_n = A_s f_y d \left(1 - \frac{A_s f_y}{1.7 f_c' bd}\right) \qquad 50.30$$

$$\rho = \frac{A_s}{bd} \qquad 50.31$$

$$M_n = \rho bd^2 f_y \left(1 - \frac{\rho f_y}{1.7 f_c'}\right)$$
$$= A_s f_y \left(d - \frac{A_s f_y}{1.7 f_c' b}\right) \qquad 50.32$$

$$\rho_{sb} = \left(\frac{0.85 \beta_1 f_c'}{f_y}\right) \left(\frac{87,000}{87,000 + f_y}\right)$$
$$50.33(b)$$

$$\rho_{\min} = \frac{3\sqrt{f_c'}}{f_y} \geq \frac{200}{f_y} \qquad 50.34(b)$$

Table 50.3 *Total Areas for Various Numbers of Bars*

bar size	nominal diameter (in)	weight (lbf/ft)	number of bars									
			1	2	3	4	5	6	7	8	9	10
no. 3	0.375	0.376	0.11	0.22	0.33	0.44	0.55	0.66	0.77	0.88	0.99	1.10
no. 4	0.500	0.668	0.20	0.40	0.60	0.80	1.00	1.20	1.40	1.60	1.80	2.00
no. 5	0.625	1.043	0.31	0.62	0.93	1.24	1.55	1.86	2.17	2.48	2.79	3.10
no. 6	0.750	1.502	0.44	0.88	1.32	1.76	2.20	2.64	3.08	3.52	3.96	4.40
no. 7	0.875	2.044	0.60	1.20	1.80	2.40	3.00	3.60	4.20	4.80	5.40	6.00
no. 8	1.000	2.670	0.79	1.58	2.37	3.16	3.95	4.74	5.53	6.32	7.11	7.90
no. 9	1.128	3.400	1.00	2.00	3.00	4.00	5.00	6.00	7.00	8.00	9.00	10.0
no. 10	1.270	4.303	1.27	2.54	3.81	5.08	6.35	7.62	8.89	10.16	11.43	12.70
no. 11	1.410	5.313	1.56	3.12	4.68	6.24	7.80	9.36	10.92	12.48	14.04	15.60
no. 14[a]	1.693	7.650	2.25	4.50	6.75	9.00	11.25	13.5	15.75	18.00	20.25	22.50
no. 18[a]	2.257	13.60	4.00	8.00	12.0	16.0	20.00	24.0	28.00	32.00	36.00	40.00

(Multiply in^2 by 645 to obtain mm^2.)
[a] No. 14 and no. 18 bars are typically used in columns only.

13. SERVICEABILITY: CRACKING

$$s_{\max} = 15\left(\frac{40{,}000}{f_s}\right) - 2.5c_c \le 12\left(\frac{40{,}000}{f_s}\right) \quad\textit{50.36}$$

14. CRACKED MOMENT OF INERTIA

Figure 50.7 *Parameters for Cracking Calculation*

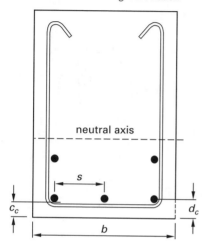

$$n = \frac{E_s}{E_c} \quad\textit{50.37}$$

$$E_c = 57{,}000\sqrt{f_c'} \quad\textit{50.38(b)}$$

$$\frac{bc_s^2}{2} = nA_s(d - c_s) \quad\textit{50.39}$$

$$c_s = \left(\frac{nA_s}{b}\right)\left(\sqrt{1 + \frac{2bd}{nA_s}} - 1\right) \quad\textit{50.40}$$

$$c_s = n\rho d\left(\sqrt{1 + \frac{2}{n\rho}} - 1\right) \quad\textit{50.41}$$

$$I_{\mathrm{cr}} = \frac{bc_s^3}{3} + nA_s(d - c_s)^2 \quad\textit{50.42}$$

Figure 50.8 *Parameters for Cracked Moment of Inertia*

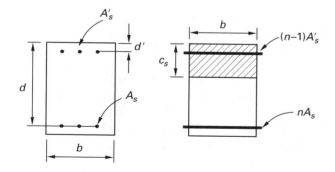

(a) reinforced section   (b) transformed section

15. SERVICEABILITY: DEFLECTIONS

$$I_e = \left(\frac{M_{\mathrm{cr}}}{M_a}\right)^3 I_g + \left(1 - \left(\frac{M_{\mathrm{cr}}}{M_a}\right)^3\right) I_{\mathrm{cr}} \le I_g \quad\textit{50.43}$$

$$M_{\mathrm{cr}} = \frac{f_r I_g}{y_t} \quad\textit{50.44}$$

$$f_r = 7.5\sqrt{f_c'} \quad\textit{50.45(b)}$$

16. LONG-TERM DEFLECTIONS

$$\Delta_a = \lambda\Delta_i \quad\textit{50.46}$$

$$\lambda = \frac{i}{1 + 50\rho'} \quad\textit{50.47}$$

17. MINIMUM BEAM DEPTHS TO AVOID EXPLICIT DEFLECTION CALCULATIONS

Table 50.4 *Minimum Beam Thickness Unless Deflections are Computed* $(f_y = 60{,}000 \text{ lbf/in}^2)$ [a]

construction	minimum h (fraction of span length)
simply supported	$\frac{1}{16}$
one end continuous	$\frac{1}{18.5}$
both ends continuous	$\frac{1}{21}$
cantilever	$\frac{1}{8}$

[a] Corrections are required for other steel strengths. See Eqs. 50.48 and 50.49.

Source: ACI 318 Table 9.5(a)

The table values are multiplied by the adjustment factor in Eq. 50.48(b) for other steel strengths.

$$0.4 + \frac{f_y}{100{,}000} \quad\textit{50.48(b)}$$

For structural lightweight concrete with a unit weight, w, between 90 and 120 lbf/ft^3, multiply the table values by Eq. 50.49(b).

$$1.65 - 0.005w \ge 1.09 \quad\textit{50.49(b)}$$

19. DESIGN OF T-BEAMS

Case 1: Beams with flanges on each side of the web [ACI 318 8.10.2]

Effective width (including the compression area of the stem) for a T-beam is the minimum of

> one fourth of the beam's span length, or

> the stem width plus 16 times the thickness of the slab, or

> the beam spacing

Case 2: Beams with an L-shaped flange [ACI 318 8.10.3]

Effective width (including the compression area of the stem) is the minimum of

> the stem width plus one twelfth of the beam's span length, or

> the stem width plus 6 times the thickness of the slab, or

> the stem width plus one half of the clear distance between beam webs

21. SHEAR REINFORCEMENT

$$\phi V_n = V_u \quad\quad 50.50$$

$$V_n = V_c + V_s \quad\quad 50.51$$

22. SHEAR STRENGTH PROVIDED BY CONCRETE

$$V_c = \left(1.9\sqrt{f_c'} + 2500\rho_w\left(\frac{V_u d}{M_u}\right)\right)b_w d \le 3.5\sqrt{f_c'}b_w d$$

$$50.52(b)$$

$$V_c = 2\sqrt{f_c'}b_w d \quad\quad 50.53(b)$$

23. SHEAR STRENGTH PROVIDED BY SHEAR REINFORCEMENT

$$V_s = \frac{A_v f_{yt} d}{s} \quad\quad 50.54$$

$$V_s = \frac{A_v f_{yt}(\sin\theta + \cos\theta)d}{s} \quad\quad 50.55$$

$$V_s = A_v f_y \sin\theta \le 3\sqrt{f_c'}b_w d \quad\quad 50.56(b)$$

24. SHEAR REINFORCEMENT LIMITATIONS

$$V_s \le 8\sqrt{f_c'}b_w d \quad\quad 50.57(b)$$

$$V_{u,max} = 10\phi\sqrt{f_c'}b_w d \quad\quad 50.58(b)$$

25. STIRRUP SPACING

$$s_{max} = \min\{24 \text{ in or } d/2\} \quad [V_s \le 4\sqrt{f_c'}b_w d]$$

$$50.59(b)$$

$$s_{max} = \min\{12 \text{ in or } d/4\} \quad [V_s > 4\sqrt{f_c'}b_w d]$$

$$50.60(b)$$

26. NO-STIRRUP CONDITIONS

$$\frac{\phi V_c}{2} \ge V_u \quad\quad 50.62$$

27. SHEAR REINFORCEMENT DESIGN PROCEDURE

$$V_{s,req} = \frac{V_u}{\phi} - V_c \quad\quad 50.63$$

$$s = \frac{A_v f_{yt} d}{V_{s,req}} \quad\quad 50.64$$

28. ANCHORAGE OF SHEAR REINFORCEMENT

$$\text{embedment length} \ge \frac{0.014 d_b f_{yt}}{\sqrt{f_c'}} \quad\quad 50.65(b)$$

30. STRENGTH ANALYSIS OF DOUBLY REINFORCED SECTIONS

$$A_c = \frac{f_y A_s - f_s' A_s'}{0.85 f_c'} \quad\quad 50.66$$

$$\varepsilon_s' = \left(\frac{0.003}{c}\right)(c - d') \quad\quad 50.67$$

$$f_s' = E_s \varepsilon_s' \le f_y \quad\quad 50.68$$

$$M_n = (A_s - A_s')f_y(d - \lambda) + A_s' f_s'(d - d') \quad 50.69$$

31. DESIGN OF DOUBLY REINFORCED SECTIONS

$$M_n' = \frac{M_u}{\phi} - M_{nc} \quad\quad 50.70$$

$$A_{s,add} = \frac{M_n'}{f_y(d - d')} \quad\quad 50.71$$

$$A_s = A_{s,max} + A_{s,add} \quad\quad 50.72$$

$$A_s' = \left(\frac{f_y}{f_s'}\right)A_{s,add} \quad\quad 50.73$$

CERM Chapter 51
Reinforced Concrete: Slabs

Chapter, section, equation, figure, and table numbers correspond to CERM. For additional study material, go to the corresponding chapter and section number in CERM.

3. TEMPERATURE STEEL

$$\rho_t = 0.0018 \left(\frac{60,000}{f_y} \right) \qquad 51.1(b)$$

6. SLAB DESIGN FOR FLEXURE

$$s = \frac{A_b}{A_{sr}} (12 \text{ in}) \qquad 51.2$$

9. DIRECT DESIGN METHOD

$$M_o = \frac{w_u l_2 l_n^2}{8} \qquad 51.3$$

10. FACTORED MOMENTS IN SLAB BEAMS

$$\alpha = \frac{E_{cb} I_b}{E_{cs} I_s} \qquad 51.4$$

Figure 51.4 *Two-Way Slab Beams (monolithic or fully composite construction) (ACI 318 13.2.4)*

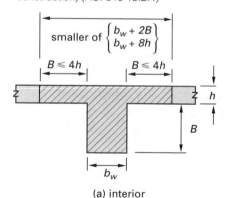

(a) interior

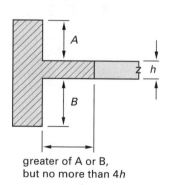

(b) exterior

11. COMPUTATION OF β_t

$$\beta_t = \frac{E_{cb} C}{2 E_{cs} I_s} \qquad 51.5$$

$$C = \sum \left(1 - 0.63 \left(\frac{x}{y} \right) \right) \left(\frac{x^3 y}{3} \right) \qquad 51.6$$

12. DEFLECTIONS IN TWO-WAY SLABS

$$t = \frac{l_n \left(0.8 + \dfrac{f_y}{200,000} \right)}{36 + 5\beta(\alpha_m - 0.2)}$$
$$\geq 5 \text{ in} \qquad 51.7$$

$$t = \frac{l_n \left(0.8 + \dfrac{f_y}{200,000} \right)}{36 + 9\beta}$$
$$\geq 3.5 \text{ in} \qquad 51.8$$

CERM Chapter 52
Reinforced Concrete: Short Columns

Chapter, section, equation, figure, and table numbers correspond to CERM. For additional study material, go to the corresponding chapter and section number in CERM.

1. INTRODUCTION

$$\frac{k_b l_u}{r} \leq 34 - 12 \left(\frac{M_1}{M_2} \right) \leq 40 \qquad 52.1$$

$$\frac{k_u l_u}{r} \leq 22 \quad [k_u > 1] \qquad 52.2$$

2. TIED COLUMNS

$$0.01 \leq \rho_g \leq 0.08 \qquad 52.3$$

3. SPIRAL COLUMNS

$$\rho_s = 0.45 \left(\frac{A_g}{A_c} - 1 \right) \left(\frac{f_c'}{f_{yt}} \right) \qquad 52.4$$

$$s \approx \frac{4 A_{sp}}{\rho_s D_c} \qquad 52.5$$

$$\text{clear distance} = s - d_{sp} \qquad 52.6$$

4. DESIGN FOR SMALL ECCENTRICITY

$$\phi\beta P_o \geq P_u \qquad 52.7$$

$$P_o = 0.85 f'_c(A_g - A_{st}) + f_y A_{st} \qquad 52.8$$

$$P_u = 1.4(\text{dead load axial force}) \\ + 1.7(\text{live load axial force}) \qquad 52.9(a)$$

$$P_u = 1.2(\text{dead load axial force}) \\ + 1.6(\text{live load axial force}) \qquad 52.9(b)$$

$$0.85 f'_c(A_g - A_{st}) + A_{st} f_y = \frac{P_u}{\phi\beta} \qquad 52.10$$

CERM Chapter 53
Reinforced Concrete: Long Columns

> Chapter, section, equation, figure, and table numbers correspond to CERM. For additional study material, go to the corresponding chapter and section number in CERM.

2. BRACED AND UNBRACED COLUMNS

$$Q = \frac{\sum P_u \Delta_o}{V_{us} l_c} \qquad 53.1$$

3. EFFECTIVE LENGTH

$$\text{effective length} = kl \qquad 53.2$$

$$\Psi = \frac{\sum_{\text{columns}} \dfrac{EI}{l_c}}{\sum_{\text{beams}} \dfrac{EI}{l_b}} \qquad 53.3$$

6. BUCKLING LOAD

$$P_c = \frac{\pi^2 EI}{(kl_u)^2} \qquad 53.6$$

7. COLUMNS IN BRACED STRUCTURES (NON-SWAY FRAMES)

$$M_c = \delta_{ns} M_2 \qquad 53.7$$

$$\delta_{ns} = \frac{C_m}{1 - \dfrac{P_u}{0.75 P_c}} \geq 1.0 \qquad 53.8$$

$$C_m = 0.6 + 0.4\left(\frac{M_1}{M_2}\right) \geq 0.4 \qquad 53.9$$

$$M_{2,\text{min,in-lbf}} = P_{u,\text{lbf}}(0.6 + 0.03 h_{\text{in}}) \qquad 53.10$$

8. COLUMNS IN UNBRACED STRUCTURES (SWAY FRAMES)

$$M_1 = M_{1,\text{ns}} + \delta_s M_{1s} \qquad 53.11$$

$$M_2 = M_{2,\text{ns}} + \delta_s M_{2s} \qquad 53.12$$

$$\delta_s = \frac{1}{1 - \dfrac{\sum P_u}{0.75 \sum P_c}} \geq 1.0 \qquad 53.13$$

$$\delta_s = \frac{1}{1 - Q} \geq 1 \qquad 53.14$$

$$\frac{l_u}{r} \geq \frac{35}{\sqrt{\dfrac{P_u}{f'_c A_g}}} \qquad 53.15$$

CERM Chapter 54
Reinforced Concrete: Walls and Retaining Walls

> Chapter, section, equation, figure, and table numbers correspond to CERM. For additional study material, go to the corresponding chapter and section number in CERM.

2. BEARING WALLS: EMPIRICAL METHOD

$$P_u \leq \phi P_{n,w} \leq 0.55 \phi f'_c A_g \left(1 - \left(\frac{kl_c}{32h}\right)^2\right) \qquad 54.1$$

5. DESIGN OF RETAINING WALLS

Figure 54.1 *Retaining Wall Dimensions (step 1)*

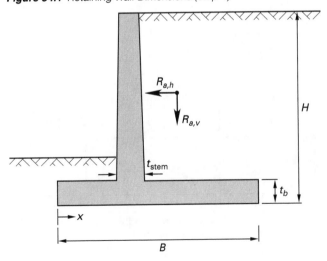

$$\rho \approx 0.02 \qquad \text{54.2}$$

$$R_{u,\text{trial}} = \rho f_y \left(1 - \left(\frac{\rho f_y}{(2)(0.85)f'_c} \right) \right) \qquad \text{54.3}$$

$$M_{u,\text{stem}} = 1.7 R_{a,h} y_a \qquad \text{54.4(a)}$$

$$M_{u,\text{stem}} = 1.6 R_{a,h} y_a \qquad \text{54.4(b)}$$

Figure 54.2 *Base of Stem Details (step 4)*

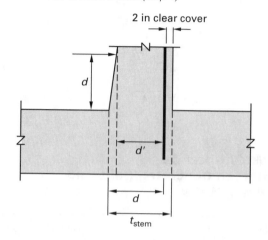

2 in clear cover

d

d'

d

t_{stem}

$$d = \sqrt{\frac{M_{u,\text{stem}}}{\phi R_u w}} \qquad \text{54.5}$$

$$t_{\text{stem}} = d + \text{cover} + \tfrac{1}{2} d_b \qquad \text{54.6}$$

$$V_{u,\text{stem}} = 1.7 V_{\text{active}} + 1.7 V_{\text{surcharge}} \qquad \text{54.7(a)}$$

$$V_{u,\text{stem}} = 1.6 V_{\text{active}} + 1.6 V_{\text{surcharge}} \qquad \text{54.7(b)}$$

$$\phi V_n = 2 \phi w d' \sqrt{f'_c} \qquad \text{54.8}$$

Figure 54.3 *Heel Details (step 9)*

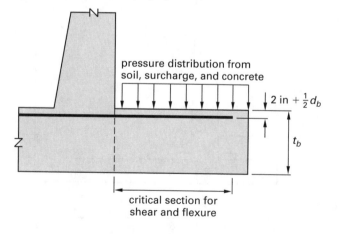

pressure distribution from
soil, surcharge, and concrete

2 in + $\tfrac{1}{2} d_b$

t_b

critical section for
shear and flexure

$$V_{u,\text{base}} = 1.4 V_{\text{soil}} + 1.4 V_{\text{heel weight}}$$
$$+ 1.7 V_{\text{surcharge}} \qquad \text{54.9(a)}$$

$$V_{u,\text{base}} = 1.2 V_{\text{soil}} + 1.2 V_{\text{heel weight}}$$
$$+ 1.6 V_{\text{surcharge}} \qquad \text{54.9(b)}$$

$$M_{u,\text{base}} = 1.4 M_{\text{soil}} + 1.4 M_{\text{heel weight}}$$
$$+ 1.7 M_{\text{surcharge}} \qquad \text{54.10(a)}$$

$$M_{u,\text{base}} = 1.2 M_{\text{soil}} + 1.2 M_{\text{heel weight}}$$
$$+ 1.6 M_{\text{surcharge}} \qquad \text{54.10(b)}$$

$$V_{u,\text{toe}} = 1.7 V_{\text{toe pressure}} - 1.4 V_{\text{toe weight}}$$
$$\text{54.11(a)}$$

$$V_{u,\text{toe}} = 1.6 V_{\text{toe pressure}} - 1.2 V_{\text{toe weight}}$$
$$\text{54.11(b)}$$

$$M_{u,\text{toe}} = 1.7 M_{\text{toe pressure}} - 1.4 M_{\text{toe weight}}$$
$$\text{54.12(a)}$$

$$M_{u,\text{toe}} = 1.6 M_{\text{toe pressure}} - 1.2 M_{\text{toe weight}}$$
$$\text{54.12(b)}$$

$$M_n = \frac{M_u}{\phi} = \rho f_y \left(1 - \frac{\rho f_y}{(2)(0.85)f'_c} \right) b d^2 \qquad \text{54.13}$$

CERM Chapter 55
Reinforced Concrete: Footings

> Chapter, section, equation, figure, and table numbers correspond to CERM. For additional study material, go to the corresponding chapter and section number in CERM.

1. INTRODUCTION

$$q_s = \frac{P_s}{A_f} + \gamma_c h + \gamma_s (H - h) \pm \frac{M_s \left(\dfrac{B}{2} \right)}{I_f} \leq q_a \quad \text{55.1}$$

$$I_f = \tfrac{1}{12} L B^3 \qquad \text{55.2}$$

2. WALL FOOTINGS

$$q_u = \frac{P_u}{B} \qquad 55.3$$

$$P_u = 1.4P_d + 1.7P_l \qquad 55.4(a)$$

$$P_u = 1.2P_d + 1.6P_l \qquad 55.4(b)$$

$$\phi v_c \geq v_u \qquad 55.5$$

$$v_c = 2\sqrt{f'_c} \qquad 55.6$$

$$v_u = \left(\frac{q_u}{d}\right)\left(\frac{B-t}{2} - d\right) \qquad 55.7$$

$$d = \frac{q_u(B-t)}{2(q_u + \phi v_c)} \qquad 55.8$$

$$h = d + \tfrac{1}{2}(\text{diameter of } x \text{ bars})$$
$$+ \text{ diameter of } z \text{ bars} + \text{cover}$$
$$\begin{bmatrix} \text{orthogonal reinforcement} \\ \text{under main steel} \end{bmatrix} \quad 55.9(a)$$

$$h = d + \tfrac{1}{2}(\text{diameter of } x \text{ bars}) + \text{cover}$$
$$\begin{bmatrix} \text{orthogonal reinforcement} \\ \text{over main steel} \end{bmatrix} \quad 55.9(b)$$

$$h \geq 6 \text{ in} + \text{diameter of } x \text{ bars}$$
$$+ \text{ diameter of } z \text{ bars} + 3 \text{ in} \qquad 55.10$$

3. COLUMN FOOTINGS

$$v_u = \frac{q_u e}{d} \qquad 55.11$$

Figure 55.3 *Critical Section for One-Way Shear*

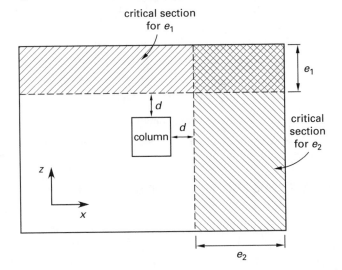

$$A_p = 2(b_1 + b_2)d \qquad 55.12$$

Figure 55.4 *Critical Section for Two-Way Shear*

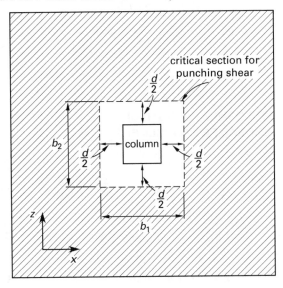

$$v_u = \frac{P_u - R}{A_p} + \frac{\gamma_v M_u(0.5b_1)}{J} \qquad 55.13$$

$$R = \frac{P_u b_1 b_2}{A_f} \qquad 55.14$$

$$\gamma_v = 1 - \frac{1}{1 + \frac{2}{3}\sqrt{\frac{b_1}{b_2}}} \qquad 55.15$$

$$J = \left(\frac{db_1^3}{6}\right)\left(1 + \left(\frac{d}{b_1}\right)^2 + 3\left(\frac{b_2}{b_1}\right)\right) \qquad 55.16$$

$$v_c = (2+y)\sqrt{f'_c} \qquad 55.17$$

$$y = \min\{2, \ 4/\beta_c, \ 40d/b_o\} \qquad 55.18$$

$$\beta_c = \frac{\text{column long side}}{\text{column short side}} \qquad 55.19$$

$$b_o = \frac{A_p}{d} = 2(b_1 + b_2) \qquad 55.20$$

$$h = d + \tfrac{1}{2}(\text{diameter of } x \text{ bars}$$
$$+ \text{ diameter of } z \text{ bars}) + \text{cover} \qquad 55.21$$

$$h \geq 6 \text{ in} + \text{diameter of } x \text{ bars}$$
$$+ \text{ diameter of } z \text{ bars} + 3 \text{ in} \qquad 55.22$$

4. SELECTION OF FLEXURAL REINFORCEMENT

$$M_u = \frac{q_u L l^2}{2} \quad [\text{no column moment}] \qquad 55.23$$

$$A_s = \frac{M_u}{\phi f_y (d - \lambda)} \qquad 55.24$$

$$A_1 = A_{sd} \left(\frac{2}{\beta + 1} \right) \qquad 55.25$$

$$A_2 = A_{sd} - A_1 \qquad 55.26$$

5. DEVELOPMENT LENGTH OF FLEXURAL REINFORCEMENT

$$l_d = \frac{3 d_b f_y \psi_t \psi_e \lambda}{50 \sqrt{f_c'}} \geq 12 \text{ in} \quad \begin{bmatrix} \text{no. 6 bars} \\ \text{and smaller} \end{bmatrix} \qquad 55.27$$

$$l_d = \frac{3 d_b f_y \psi_t \psi_e \lambda}{40 \sqrt{f_c'}} \geq 12 \text{ in} \quad [\text{no. 7 bars and larger}] \qquad 55.28$$

$$l_{dh} = \frac{0.02 \psi_e d_b f_y}{\sqrt{f_c'}} \geq 8 d_b \geq 6 \text{ in} \qquad 55.29$$

6. TRANSFER OF FORCE AT COLUMN BASE

$$A_{db,\min} = 0.005 A_g \qquad 55.30$$

$$l_{dc} = 0.02 d_b \left(\frac{f_y}{\sqrt{f_c'}} \right)$$
$$\geq 0.0003 d_b f_y \geq 8 \text{ in} \qquad 55.31$$

$$l_d \leq h - \text{cover} - \text{diameter of } x \text{ steel}$$
$$- \text{diameter of } z \text{ steel} \qquad 55.32$$

$$P_{\text{bearing,column}} = 0.85 \phi f_{c,\text{column}}' A_c \qquad 55.33$$

$$P_{\text{bearing,footing}} = 0.85 \alpha \phi f_{c,\text{footing}}' A_c \qquad 55.34$$

$$\alpha = \sqrt{\frac{A_{ff}}{A_c}} \leq 2 \qquad 55.35$$

$$A_{db} = \frac{P_u - \text{smaller of} \begin{Bmatrix} P_{\text{bearing,column}} \\ P_{\text{bearing,footing}} \end{Bmatrix}}{\phi f_y}$$
$$\geq A_{db,\min} \qquad 55.36$$

CERM Chapter 56
Pretensioned Concrete

Chapter, section, equation, figure, and table numbers correspond to CERM. For additional study material, go to the corresponding chapter and section number in CERM.

7. CREEP AND SHRINKAGE

$$\Delta_{\text{long term}} = \lambda \Delta_{\text{immediate}} \qquad 56.1$$

8. PRESTRESS LOSSES

$$n = \frac{E_s}{E_c} \qquad 56.2$$

9. DEFLECTIONS

Figure 56.2 *Midspan Deflections from Prestressing*[a]

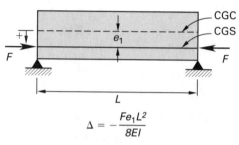

$$\Delta = -\frac{F e_1 L^2}{8EI}$$

(a) straight tendons

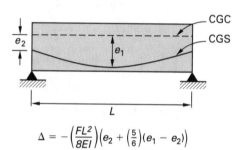

$$\Delta = -\left(\frac{FL^2}{8EI} \right) \left(e_2 + \left(\frac{5}{6} \right) (e_1 - e_2) \right)$$

(b) parabolically draped tendons

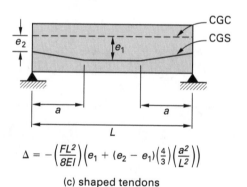

$$\Delta = -\left(\frac{FL^2}{8EI} \right) \left(e_1 + (e_2 - e_1) \left(\frac{4}{3} \right) \left(\frac{a^2}{L^2} \right) \right)$$

(c) shaped tendons

[a]CGC—centroid of concrete; CGS—centroid of prestressing tendons

13. ACI CODE PROVISIONS FOR STRENGTH

$$f_{ps} = f_{pu} \left(1 - \left(\frac{\gamma_p}{\beta_1} \right) \left(\rho_p \left(\frac{f_{pu}}{f_c'} \right) + \left(\frac{d}{d_p} \right) (\omega - \omega') \right) \right)$$
$$[\text{bonded tendons}] \qquad 56.3$$

$$\rho_p = \frac{A_{ps}}{b d_p} \qquad 56.4$$

$$\rho_p \left(\frac{f_{pu}}{f_c'} \right) + \left(\frac{d}{d_p} \right) (\omega - \omega') \geq 0.17 \qquad 56.5$$

14. ANALYSIS OF PRESTRESSED BEAMS

$$f_c = \frac{-P}{A} + \frac{Pey}{I} - \frac{M_s y}{I} \qquad 56.6$$

15. SHEAR IN PRESTRESSED SECTIONS

$$V_u < \phi V_n \qquad 56.7$$

$$V_n = V_c + V_s \qquad 56.8$$

CERM Chapter 57
Composite Concrete and
Steel Bridge Girders

> Chapter, section, equation, figure, and table numbers correspond to CERM. For additional study material, go to the corresponding chapter and section number in CERM.

5. EFFECTIVE SLAB WIDTH

$$b_e = \text{smaller of } \{L/4;\ b_o\} \qquad 57.1$$

$$b_e = \text{smallest of } \{L/4;\ 12t;\ b_o\} \qquad 57.2$$

$$b_e = \text{smallest of } \{L/12;\ 6t;\ \tfrac{1}{2}b_o\} \qquad 57.3$$

6. SECTION PROPERTIES

$$n = \frac{E_s}{E_c} \qquad 57.4$$

CERM Chapter 59
Structural Steel: Beams

> Chapter, section, equation, figure, and table numbers correspond to CERM. For additional study material, go to the corresponding chapter and section number in CERM.

3. BENDING STRENGTH IN STEEL BEAMS

$$M_n = F_y Z_x \quad \text{[AISC Eq. F2-1]} \qquad 59.1$$

$$M \leq \frac{M_n}{\Omega_b} \qquad 59.2$$

4. COMPACT SECTIONS

$$\frac{b_f}{2t_f} \leq 0.38\sqrt{\frac{E}{F_y}} \quad \begin{bmatrix} \text{flanges in flexural} \\ \text{compression only} \end{bmatrix} \qquad 59.3$$

$$\frac{h}{t_w} \leq 3.76\sqrt{\frac{E}{F_y}} \quad \begin{bmatrix} \text{webs in flexural} \\ \text{compression only} \end{bmatrix} \qquad 59.4$$

5. LATERAL BRACING

Figure 59.11 *Available Moment vs. Unbraced Length*

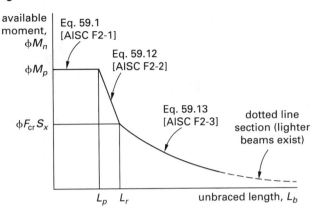

$$L_p = 1.76 r_y \sqrt{\frac{E}{F_y}} \quad \text{[AISC Eq. F2-5]} \qquad 59.5$$

$$L_r = 1.95 r_{ts} \frac{E}{0.7F_y}$$

$$\times \sqrt{\frac{Jc}{S_x h_o}}$$

$$\times \sqrt{1 + \sqrt{1 + 6.76\left(\left(\frac{0.7F_y}{E}\right)\left(\frac{S_x h_o}{Jc}\right)\right)^2}}$$

$$\text{[AISC Eq. F2-6]} \quad 59.6$$

$$r_{ts} = \frac{b_f}{\sqrt{12\left(1 + \frac{ht_w}{6b_f t_f}\right)}} \qquad 59.7$$

$$r_{ts}^2 = \frac{\sqrt{I_y C_w}}{S_x} \qquad 59.8$$

$$C_w = \frac{h^2 I_y}{4} \approx \frac{(d - t_f)^2 t_f b_f^3}{24} \qquad 59.9$$

$$J = \frac{1}{3}\int_0^b t^3\, ds \approx \frac{1}{3}\sum bt^3 \quad [b > t;\ b/t > 10] \qquad 59.10$$

6. LATERAL TORSIONAL BUCKLING

$$M_{n,\text{non-uniform moment}} = C_b M_{n,\text{uniform moment}} \qquad 59.12$$

$$C_b = \frac{12.5M_{\max}}{2.5M_{\max} + 3M_A + 4M_B + 3M_C} \times R_m \leq 3.0 \qquad 59.13$$

7. FLEXURAL DESIGN STRENGTH: I-SHAPES BENDING ABOUT MAJOR AXIS

If $L_r \geq L_b > L_p$, then

$$M_n = C_b \left(M_p - (M_p - 0.7F_yS_x) \left(\frac{L_b - L_p}{L_r - L_p} \right) \right)$$
$$\leq M_p \quad \text{[AISC Eq. F2-2]} \qquad 59.14$$

If $L_b > L_r$, or if there is no bracing at all between support points, then

$$M_n = F_{cr}S_x \leq M_p \quad \text{[AISC Eq. F2-3]} \qquad 59.15$$

$$F_{cr} = \frac{C_b\pi^2 E}{\left(\frac{L_b}{r_{ts}}\right)^2} \sqrt{1 + 0.078 \frac{Jc}{S_x h_o} \left(\frac{L_b}{r_{ts}}\right)^2}$$

$$\text{[AISC Eq. F2-4]} \qquad 59.16$$

9. SHEAR STRENGTH IN STEEL BEAMS

$$V_n = 0.6F_yA_wC_v$$

10. CONCENTRATED WEB FORCES

Figure 59.6 *Nomenclature for Web Yielding Calculations*

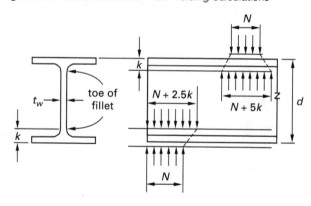

$$R_n = (5k + N)F_{yw}t_w \qquad 59.17$$

$$R_n = (2.5k + N)F_{yw}t_w \qquad 59.18$$

$$R_n = 0.80t_w^2 \left(1 + 3\left(\frac{N}{d}\right)\left(\frac{t_w}{t_f}\right)^{1.5}\right)\sqrt{\frac{EF_{yw}t_f}{t_w}}$$
$$\text{[interior loads]} \quad \text{[AISC Eq. J10-4]} \qquad 59.19$$

$$R_n = 0.40t_w^2 \left(1 + 3\left(\frac{N}{d}\right)\left(\frac{t_w}{t_f}\right)^{1.5}\right)\sqrt{\frac{EF_{yw}t_f}{t_w}}$$
$$[N/d \leq 0.2] \quad \text{[AISC Eq. J10-5a]} \qquad 59.20$$

$$R_n = 0.40t_w^2 \left(1 + \left(\frac{4N}{d} - 0.2\right)\left(\frac{t_w}{t_f}\right)^{1.5}\right)\sqrt{\frac{EF_{yw}t_f}{t_w}}$$
$$[N/d > 0.2] \quad \text{[AISC Eq. J10-5b]} \qquad 59.21$$

14. PLASTIC DESIGN OF CONTINUOUS BEAMS

$$\text{factored load} = 1.7 \times \text{dead load}$$
$$+ 1.7 \times \text{live load} \qquad 59.22$$

Figure 59.12 *Location of Plastic Moments on a Uniformly Loaded Continuous Beam (plastic hinges shown as •)*

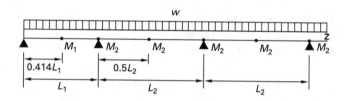

$$Z_x = \frac{M_p}{F_y} \qquad 59.23$$

15. ULTIMATE PLASTIC MOMENTS

$$M_1 = \left(\tfrac{3}{2} - \sqrt{2}\right)wL_1^2$$
$$\approx 0.0858wL_1^2 \qquad 59.26$$
$$M_2 = \frac{wL^2}{16} \qquad 59.27$$

16. ULTIMATE SHEARS

$$V_{max} = \frac{wL}{2} + \frac{M_{max}}{L} = 0.5858wL \qquad 59.28$$

18. UNSYMMETRICAL BENDING

$$\left| \frac{f_a}{F_a} + \frac{f_{by}}{F_{by}} + \frac{f_{by}}{F_{by}} \right| \leq 1.0 \qquad 59.29$$

$$\frac{f_{bx}}{F_{bx}} + \frac{f_{by}}{F_{by}} \leq 1.0 \qquad 59.30$$

21. BEAM BEARING PLATES

Figure 59.14 Nomenclature for Beam Bearing Plate

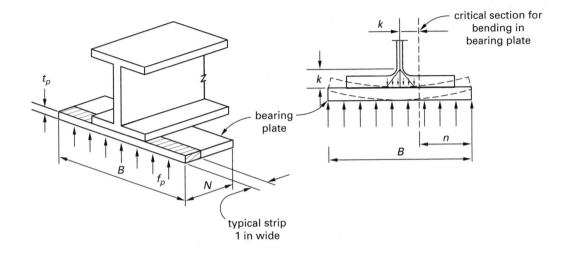

$$P_p = 0.85 f'_c A_1 \quad \text{[AISC Eq. J8-1]} \qquad 59.31$$

$$P_p = 0.85 f'_c A_1 \sqrt{\frac{A_2}{A_1}} \le 1.7 f'_c A_1 \quad \text{[AISC Eq. J8-2]} \qquad 59.32$$

$$N = \frac{\phi R_n - \phi R_1}{\phi R_2} \qquad 59.33$$

$$N = \frac{\phi R_n - \phi R_3}{\phi R_4} \qquad 59.34$$

$$B = \frac{A_1}{N} \qquad 59.35$$

$$f_p = \frac{R}{BN} \qquad 59.36$$

$$n = \frac{B}{2} - k \qquad 59.37$$

$$t = \sqrt{\frac{3 f_p n^2}{F_b}} \qquad 59.38$$

CERM Chapter 60
Structural Steel: Tension Members

Chapter, section, equation, figure, and table numbers correspond to CERM. For additional study material, go to the corresponding chapter and section number in CERM.

2. AXIAL TENSILE STRENGTH

$$P_u \le \phi_t P_n \qquad 60.1$$

3. GROSS AREA

$$A_g = bt \qquad 60.2$$

4. NET AREA

$$b_n = b - \sum d_h + \sum \frac{s^2}{4g} \qquad 60.3$$

$$A_n = b_n t \qquad 60.4$$

$$A_n = A_g - \sum d_h t + \left(\sum \frac{s^2}{4g} \right) t \qquad 60.5$$

Figure 60.4 Tension Member with Uniform Thickness and Unstaggered Holes

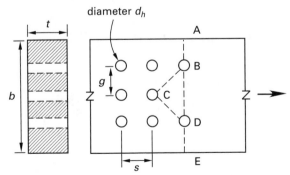

5. EFFECTIVE NET AREA

$$A_e = U A_n \qquad 60.6$$

Table 60.1 *Reduction Coefficient (U) Values for Bolted Connections*

(a) For W, M, S, or HP shapes with flange widths not less than two thirds of the depth, and for structural tees cut from these shapes (provided the connection is to the flanges) with three or more fasteners per line in the direction of stress): $U = 0.90$.

(b) For W, M, S, or HP structural shapes not meeting the conditions of paragraph (a), structural tees cut from these shapes, and all other shapes, including built-up cross sections, with three or more fasteners per line in the direction of stress: $U = 0.85$.

(c) For all members with only two fasteners per line in the direction of stress: $U = 0.60$.

Source: *AISC Specification* Table D3.1

$$A_e = U A_g \qquad 60.7$$

Table 60.2 *Reduction Coefficient (U) Values for Welded Connections*[a]

$l \geq 2w$	$U = 1.0$
$2w > l \geq 1.5w$	$U = 0.87$
$1.5w > l \geq w$	$U = 0.75$

[a]l = weld length, in; w = plate width (distance between welds), in

Source: *AISC Specification* Table D3.1

6. DESIGN TENSILE STRENGTH

$$P_n = F_y A_g \quad \text{[yielding criterion; AISC Eq. D2-1]} \qquad 60.8$$
$$P_n = F_u A_e \quad \text{[fracture criterion; AISC Eq. D2-2]} \qquad 60.9$$

7. BLOCK SHEAR STRENGTH

$$R_n = 0.6 F_u A_{nv} + U_{bs} F_u A_{nt}$$
$$\leq 0.6 F_y A_{gv} + U_{bs} F_u A_{nt} \quad \begin{bmatrix} \text{block shear criterion;} \\ \text{AISC Eq. J4-5} \end{bmatrix}$$
$$\qquad 60.10$$

8. SLENDERNESS RATIO

$$\text{SR} = \frac{L}{r_y} \qquad 60.11$$

CERM Chapter 61
Structural Steel: Compression Members

> Chapter, section, equation, figure, and table numbers correspond to CERM. For additional study material, go to the corresponding chapter and section number in CERM.

2. EULER'S COLUMN BUCKLING THEORY

$$P_e = \frac{\pi^2 E I}{L^2} \qquad 61.1$$

$$F_e = \frac{P_e}{A} = \frac{\pi^2 E}{\left(\dfrac{L}{r}\right)^2} \qquad 61.2$$

3. EFFECTIVE LENGTH

$$F_e = \frac{\pi^2 E}{\left(\dfrac{KL}{r}\right)^2} \quad \text{[AISC Eq. E3-4]} \qquad 61.3$$

$$G = \frac{\sum \left(\dfrac{I}{L}\right)_{\text{columns}}}{\sum \left(\dfrac{I}{L}\right)_{\text{beams}}} \qquad 61.4$$

Table 61.1 *Effective Length Factors*[a]

end no. 1	end no. 2	K
built-in	built-in	0.65
built-in	pinned	0.80
built-in	rotation fixed, translation free	1.2
built-in	free	2.1
pinned	pinned	1.0
pinned	rotation fixed, translation free	2.0

[a]These are slightly different from the theoretical values often quoted for use with Euler's equation.

Source: *AISC Commentary*, Table C-C2.2

5. SLENDERNESS RATIO

$$\text{SR} = \frac{KL}{r} \qquad 61.5$$

7. DESIGN COMPRESSIVE STRENGTH

Figure 61.4 *Available Compressive Stress Versus Slenderness Ratio*

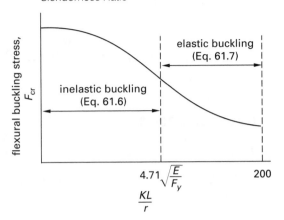

$$F_{\mathrm{cr}} = \left(0.658^{F_y/F_e}\right) F_y \qquad 61.6$$

$$F_{\mathrm{cr}} = 0.877 F_e \qquad 61.7$$

8. ANALYSIS OF COLUMNS

$$P_n = F_{\mathrm{cr}} A_g \quad \text{[AISC Eq. E3-1]} \qquad 61.8$$

$$\mathrm{SR} = \text{larger of} \left\{ \begin{array}{c} \dfrac{K_x L_x}{r_x} \\[2mm] \dfrac{K_y L_y}{r_y} \end{array} \right\} \qquad 61.9$$

$$\frac{P_n}{\Omega_c} = 1.67 F_{\mathrm{cr}} A_g \quad \text{[ASD]} \qquad 61.10(a)$$

$$\phi_c P_n = 0.90 F_{\mathrm{cr}} A_g \quad \text{[LRFD]} \qquad 61.10(b)$$

9. DESIGN OF COLUMNS

$$\text{effective length} = K_y L_y \qquad 61.11$$

$$K_x L_x' = \frac{K_x L_x}{\dfrac{r_x}{r_y}} \qquad 61.12$$

10. LOCAL BUCKLING

$$\frac{b}{t} \le \frac{H}{\sqrt{F_y}} \qquad 61.13$$

$$\frac{D}{t} \le 0.11 \left(\frac{E}{F_y}\right) \qquad 61.14$$

Table 61.2 *H Values for Width-Thickness Ratios*

element	H
unstiffened elements	
stems of tees	127
double angles in contact	95
compression flanges of beams	95
angles or plates projecting from girders, columns, or other compression members, and compression flanges of plate girders[a,b]	$109\sqrt{k_c}$
stiffeners on plate girders	95
flanges of tees and I-beams (use $b_f/2$)	95
single-angle struts or separated double-angle struts	76
stiffened elements	
square and rectangular box sections	190
other uniformly compressed elements	253

[a] $k_c = 4/\sqrt{h/t}$; $0.35 < k_c < 0.76$

[b] Other provisions govern the width-thickness ratios of plate girder flanges, as well.

Source: *AISC Specification* Table B4.1

12. COLUMN BASE PLATES

$$F_p = 0.85 f_c' \sqrt{\frac{A_2}{A_1}} \qquad 61.15$$

$$A_{\mathrm{plate}} = \frac{\text{column load}}{F_p} \qquad 61.16$$

$$f_p = \frac{\text{column load}}{\text{actual plate area}} \qquad 61.17$$

$$A_1 = \frac{\text{column load}}{F_p} \qquad 61.18$$

$$N = 2m + 0.95d \qquad 61.19$$

$$B = \frac{A_1}{N} \qquad 61.20$$

$$X = \left(\frac{4 d b_f}{(d + b_f)^2}\right)\left(\frac{P_u}{\phi_c P_p}\right) \quad \text{[LRFD]} \qquad 61.21(a)$$

$$X = \left(\frac{4 d b_f}{(d + b_f)^2}\right)\left(\frac{\Omega_c P_a}{P_p}\right) \quad \text{[ASD]} \qquad 61.21(b)$$

Figure 61.7 *Column Base Plate*

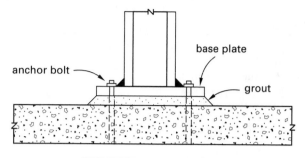

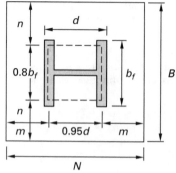

$$\lambda = \frac{2\sqrt{X}}{1 + \sqrt{1 - X}} \leq 1 \qquad 61.22$$

$$t_{\min} = l\sqrt{\frac{2P_u}{0.9 F_y BN}} \quad \text{[LRFD]} \qquad 61.23(a)$$

$$t_{\min} = l\sqrt{\frac{3.33 P_a}{F_y BN}} \quad \text{[ASD]} \qquad 61.23(b)$$

CERM Chapter 62
Structural Steel: Beam-Columns

Chapter, section, equation, figure, and table numbers correspond to CERM. For additional study material, go to the corresponding chapter and section number in CERM.

1. INTRODUCTION

$$f_{\max} = f_a + (\text{AF})_x f_{bx} + (\text{AF})_y f_{by} \qquad 62.1$$

$$\frac{f_a}{F_a} + (\text{AF})_x \left(\frac{f_{bx}}{F_{bx}}\right) + (\text{AF})_y \left(\frac{f_{by}}{F_{by}}\right) = 1.0 \qquad 62.2$$

2. FLEXURAL/AXIAL COMPRESSION

For $P_r/P_c \geq 0.2$,

$$\frac{P_r}{P_c} + \left(\frac{8}{9}\right)\left(\frac{M_{rx}}{M_{cx}} + \frac{M_{ry}}{M_{cy}}\right) \leq 1.0 \qquad 62.3$$

For $P_r/P_c < 0.2$,

$$\frac{P_r}{2P_c} + \frac{M_{rx}}{M_{cx}} + \frac{M_{ry}}{M_{cy}} \leq 1.0 \qquad 62.4$$

$$C_b = \left(\frac{12.5 M_{\max}}{2.5 M_{\max} + 3M_A + 4M_B + 3M_C}\right) R_m \leq 3.0 \qquad 62.5$$

4. DESIGN OF BEAM-COLUMNS

For $P_r/P_c \geq 0.2$,

$$pP_r + b_x M_{rx} + b_y M_{ry} \leq 1.0 \quad \text{[large axial loads]} \qquad 62.8$$

For $P_r/P_c < 0.2$,

$$\frac{pP_r}{2} + \left(\frac{9}{8}\right)(b_x M_{rx} + b_y M_{ry}) \leq 1.0 \quad \text{[small axial loads]} \qquad 62.9$$

CERM Chapter 63
Structural Steel: Plate Girders

Chapter, section, equation, figure, and table numbers correspond to CERM. For additional study material, go to the corresponding chapter and section number in CERM.

1. INTRODUCTION

Figure 63.1 *Elements of a Plate Girder*

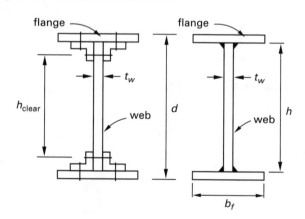

2. DEPTH-THICKNESS RATIOS

$$\left(\frac{h}{t_w}\right)_{\max} \leq 11.7 \sqrt{\frac{E}{F_y}} \qquad 63.1$$

$$\left(\frac{h}{t_w}\right)_{\max} \leq \frac{0.42 E}{F_y} \qquad 63.2$$

3. SHEAR STRENGTH

$$V_n = 0.6 F_y A_w C_v \quad \text{[AISC Eq. G2-1]} \qquad 63.3$$

$$C_v = 1 \quad \text{[AISC Eq. G2-3]} \qquad 63.4$$

$$C_v = \frac{1.10 \sqrt{\dfrac{k_v E}{F_y}}}{\dfrac{h}{t_w}} \quad \text{[AISC Eq. G2-4]} \qquad 63.5$$

$$C_v = \frac{1.51 E k_v}{\left(\dfrac{h}{t_w}\right)^2 F_y} \quad \text{[AISC Eq. G2-5]} \qquad 63.6$$

4. DESIGN OF GIRDER WEBS AND FLANGES

$$A_f = \frac{M_p}{F_y h} - \frac{ht}{4} \qquad 63.7$$

$$A_f = b_f t_f \qquad 63.8$$

5. WIDTH-THICKNESS RATIOS

$$\frac{b_f}{2t_f} \le \frac{64.7}{\sqrt{F_{yf}}} \qquad 63.9$$

$$\frac{b_f}{2t_f} \le 162 \sqrt{\frac{k_c}{F_L}} \qquad 63.10$$

$$0.35 \le \frac{4}{\sqrt{\dfrac{h}{t_w}}} \le 0.76 \qquad 63.11$$

8. LOCATION OF INTERIOR STIFFENERS

$$k_v = 5 + \frac{5}{\left(\dfrac{a}{h}\right)^2} \qquad 63.12$$

Equation 63.12 applies when $a/h > 3.0$, or

$$\frac{a}{h} > \left(\frac{260}{\dfrac{h}{t_w}}\right)^2 \qquad 63.13$$

$$V_n = 0.6 F_y A_w \left(C_v + \frac{1 - C_v}{1.15 \sqrt{1 + \left(\dfrac{a}{h}\right)^2}} \right) \quad \text{[AISC G3-2]}$$
$$63.14$$

$$V_n = 0.6 F_y A_w \qquad 63.15$$

9. DESIGN OF INTERMEDIATE STIFFENERS

$$A_{st} > \frac{F_y}{F_{y,st}} \left(0.15 D_s h t_w (1 - C_v) \frac{V_r}{V_c} - 18 t_w^2 \right) \ge 0$$
$$63.16$$

$$\left(\frac{b}{t}\right)_{st} \le 0.56 \sqrt{\frac{E}{F_{y,st}}} \qquad 63.17$$

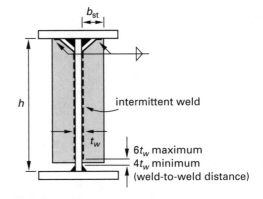

Figure 63.2 *Intermediate Stiffeners*

10. DESIGN OF BEARING STIFFENERS

$$\frac{b_{st}}{t_{st}} \le \frac{95}{\sqrt{F_y}} \qquad 63.18$$

$$\frac{l}{r} = \frac{0.75h}{0.25(2b_{st} + t_w)} \qquad 63.19$$

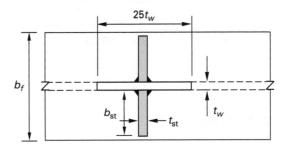

Figure 63.3 *Bearing Stiffener (top view)*

$$t_{st} = \frac{\dfrac{\text{load}}{\phi F_{cr}} - 25 t_w^2}{2 b_{st}} \qquad 63.20$$

$$t_{st} = \frac{\dfrac{\text{load}}{0.9 F_y}}{2 b_{st}} \qquad 63.21$$

CERM Chapter 64
Structural Steel: Composite Beams

Chapter, section, equation, figure, and table numbers correspond to CERM. For additional study material, go to the corresponding chapter and section number in CERM.

2. EFFECTIVE WIDTH OF CONCRETE SLAB

$$b = \text{smaller of} \begin{Bmatrix} \dfrac{L}{4} \\ s \end{Bmatrix} \quad \text{[interior beams]} \qquad 64.1$$

3. SECTION PROPERTIES

$$I_{nb} = I_s + A_s \left(Y_{\text{ENA}} - \frac{d}{2} \right)^2$$
$$+ \left(\frac{\sum Q_n}{F_y} \right) (d + Y_2 - Y_{\text{ENA}})^2 \qquad 64.2$$

Figure 64.2 Deflection Design Model for Composite Beams

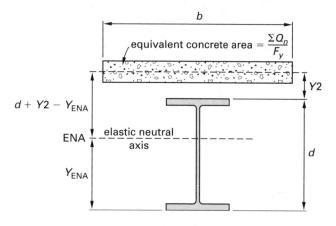

4. AVAILABLE FLEXURAL STRENGTH

$$M_n = A_g F_y \left(\frac{d}{2} \right) \qquad 64.3$$

5. SHEAR CONNECTORS

$$N_1 = \frac{A_g F_y}{Q_n} \quad \text{[full composite action]} \qquad 64.4$$

$$N_1 = \frac{C_c}{Q_n} \quad \text{[partial composite action]} \qquad 64.5$$

$$N = \frac{\sum Q_n}{Q_n} \qquad 64.6$$

CERM Chapter 65
Structural Steel: Connectors

Chapter, section, equation, figure, and table numbers correspond to CERM. For additional study material, go to the corresponding chapter and section number in CERM.

4. AVAILABLE LOADS FOR FASTENERS

Table 65.1 Nominal Fastener Stresses for Static Loading

type of connector	F_{nt} (ksi)	F_{nv} (ksi) bearing
A307 common bolts	45	24
A325 high-strength bolts		
no threads in shear plane	90	60
threads in shear plane	90	48
A490 high-strength bolts		
no threads in shear plane	113	75
threads in shear plane	113	60

Source: Based on *AISC Specification* Table J3.2.

$$R_n = \mu D_u h_{\text{sc}} T_b N_s \quad \begin{bmatrix} \text{slip resistance,} \\ \phi R_n \text{ or } \dfrac{R_n}{\Omega} \end{bmatrix} \qquad 65.1$$

$$T_{b,\min} = 0.70 A_{b,\text{th}} F_u \quad \text{[rounded]} \qquad 65.2$$

5. AVAILABLE BEARING STRENGTH

$$R_n = F_n A_b \qquad 65.3$$

8. ULTIMATE STRENGTH OF ECCENTRIC SHEAR CONNECTIONS

$$P = C R_n \qquad 65.4$$

9. TENSION CONNECTIONS

$$P = A_b F_{nt} \qquad 65.5$$

10. COMBINED SHEAR AND TENSION CONNECTIONS

Figure 65.4 Combined Shear and Tension Connection

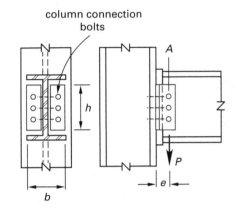

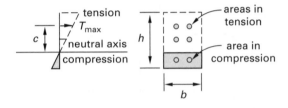

CERM Chapter 66
Structural Steel: Welding

Chapter, section, equation, figure, and table numbers correspond to CERM. For additional study material, go to the corresponding chapter and section number in CERM.

3. FILLET WELDS

Figure 66.4 Fillet Weld

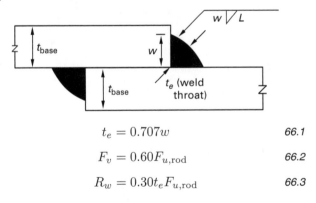

$$t_e = 0.707w \qquad 66.1$$

$$F_v = 0.60F_{u,\text{rod}} \qquad 66.2$$

$$R_w = 0.30t_eF_{u,\text{rod}} \qquad 66.3$$

Table 66.1 Minimum Fillet Weld Size

thickness of thinnest part joined (in)	minimum w (in)
to $\frac{1}{4}$ inclusive	$\frac{1}{8}$
over $\frac{1}{4}$ to $\frac{1}{2}$ inclusive	$\frac{3}{16}$
over $\frac{1}{2}$ to $\frac{3}{4}$ inclusive	$\frac{1}{4}$
over $\frac{3}{4}$	$\frac{5}{16}$

From *AISC Specification* Table J2.4.

4. CONCENTRIC TENSION CONNECTIONS

$$f_v = \frac{P}{A_{\text{weld}}} = \frac{P}{l_{\text{weld}}t_e} \qquad 66.4$$

7. ELASTIC METHOD FOR ECCENTRIC SHEAR/TORSION CONNECTIONS

$$I_p = I_x + I_y \qquad 66.10$$

$$r_m = \frac{Mc}{I_p} = \frac{Pec}{I_p} \qquad 66.11$$

$$r_p = \frac{P}{A} \qquad 66.12$$

$$r_a = \sqrt{(r_{m,y} + r_p)^2 + (r_{m,x})^2} \qquad 66.13$$

8. AISC INSTANTANEOUS CENTER OF ROTATION METHOD FOR ECCENTRIC SHEAR/TORSION CONNECTIONS

$$R_n = CC_1Dl \qquad 66.14$$

9. ELASTIC METHOD FOR OUT OF PLANE LOADING

Figure 66.7 Welded Connection in Combined Shear and Bending

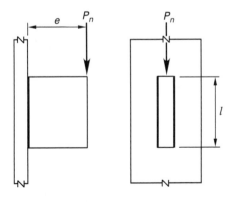

The nominal shear stress is

$$f_v = \frac{P_n}{A_{\text{weld}}} = \frac{P_n}{2Lt_e} \qquad 66.15$$

$$f_b = \frac{Mc}{I} = \frac{M}{S} = \frac{P_n e}{S} \qquad 66.16$$

$$f = \sqrt{f_v^2 + f_b^2} \qquad 66.17$$

CERM Chapter 67
Properties of Masonry

Chapter, section, equation, figure, and table numbers correspond to CERM. For additional study material, go to the corresponding chapter and section number in CERM.

7. MODULUS OF ELASTICITY

$$E_m = 700 f'_m \quad \text{[clay masonry]} \qquad 67.1$$

$$E_m = 900 f'_m \quad \text{[concrete masonry]} \qquad 67.2$$

17. DEVELOPMENT LENGTH

$$l_d = 0.0015 d_b F_s \quad \text{[wires in tension]} \qquad 67.3(b)$$

$$l_d = \frac{0.13 d_b^2 f_y \gamma}{K \sqrt{f'_m}} \quad \text{[reinforcing bars]} \qquad 67.4(b)$$

Increase lengths by 50% for epoxy-coated bars and wires.

CERM Chapter 68
Masonry Walls

Chapter, section, equation, figure, and table numbers correspond to CERM. For additional study material, go to the corresponding chapter and section number in CERM.

8. ASD WALL DESIGN: FLEXURE— REINFORCED

Figure 68.1 Stress Distribution for Fully Grouted Masonry

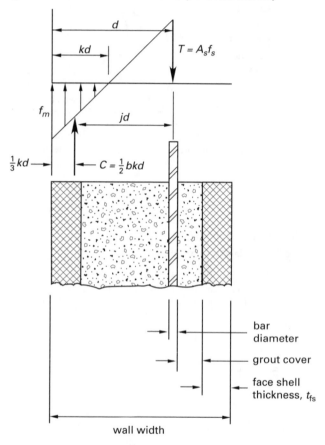

$$F_b = \tfrac{1}{3} f'_m \qquad 68.1$$

$$F_v = \sqrt{f'_m} \qquad 68.2$$

$$\rho = \frac{A_s}{bd} \qquad 68.3$$

$$k = \sqrt{2\rho n + (\rho n)^2} - \rho n \qquad 68.4$$

$$j = 1 - \frac{k}{3} \qquad 68.5$$

$$M_m = F_b bd^2 \left(\frac{jk}{2}\right) \qquad 68.6$$

$$M_s = A_s F_s jd \qquad 68.7$$

$$M_R = \text{the lesser of } M_m \text{ and } M_s \qquad 68.8$$

$$V_R = F_v bd \qquad 68.9$$

$$\rho_{\text{bal}} = \frac{nF_b}{2F_s\left(n + \dfrac{F_s}{F_b}\right)} \quad \text{[balanced]} \tag{68.10}$$

$$k = \frac{-A_s n - t_{\text{fs}}(b - b_w)}{db_w}$$
$$+ \frac{\sqrt{\begin{array}{c}(A_s n + t_{\text{fs}}(b - b_w))^2 \\ + t_{\text{fs}}^2 b_w(b - b_w) + 2db_w A_s n\end{array}}}{db_w} \tag{68.11}$$

$$j = \left(\frac{1}{kdb_w + t_{\text{fs}}(b - b_w)\left(2 - \dfrac{t_{\text{fs}}}{kd}\right)}\right)$$
$$\times \left(\begin{array}{c} kb_w\left(d - \dfrac{kd}{3}\right) + \left(\dfrac{2t_{\text{fs}}(b - b_w)}{kd^2}\right) \\ \times \left(\begin{array}{c} (kd - t_{\text{fs}})\left(d - \dfrac{t_{\text{fs}}}{2}\right) \\ + \left(\dfrac{t_{\text{fs}}}{2}\right)\left(d - \dfrac{t_{\text{fs}}}{3}\right) \end{array}\right) \end{array}\right) \tag{68.12}$$

$$M_m = \tfrac{1}{2}F_b kdb_w\left(d - \frac{kd}{3}\right) + F_b t_{\text{fs}}(b - b_w)$$
$$\times \left(\begin{array}{c} \left(1 - \dfrac{t_{\text{fs}}}{kd}\right)\left(d - \dfrac{t_{\text{fs}}}{2}\right) \\ + \left(\dfrac{t_{\text{fs}}}{2kd}\right)\left(d - \dfrac{t_{\text{fs}}}{3}\right) \end{array}\right) \tag{68.13}$$

$$M_s = A_s F_s jd \tag{68.14}$$

$$M_R = \text{the lesser of } M_m \text{ and } M_s \tag{68.15}$$

$$V_R = F_v\left(bt_{\text{fs}} + b_w(d - t_{\text{fs}})\right) \tag{68.16}$$

Table 68.3 *Reinforcing Steel Areas, A_s, in in^2/ft*

bar spacing (in)	reinforcing bar size						
	no. 3	no. 4	no. 5	no. 6	no. 7	no. 8	no. 9
8	0.166	0.295	0.460	0.663	0.902	1.178	1.491
16	0.083	0.147	0.230	0.331	0.451	0.589	0.746
24	0.055	0.098	0.153	0.221	0.301	0.393	0.497
32	0.041	0.074	0.115	0.166	0.225	0.295	0.373
40	0.033	0.059	0.092	0.133	0.180	0.236	0.298
48	0.028	0.049	0.077	0.110	0.150	0.196	0.249
56	0.024	0.042	0.066	0.095	0.129	0.168	0.213
64	0.021	0.037	0.058	0.083	0.113	0.147	0.186
72	0.018	0.033	0.051	0.074	0.100	0.131	0.166

9. ASD WALL DESIGN: AXIAL COMPRESSION AND FLEXURE—UNREINFORCED

$$\frac{f_a}{F_a} + \frac{f_b}{F_b} \leq 1 \tag{68.17}$$

$$P \leq \tfrac{1}{4}P_e \tag{68.18}$$

$$P_e = \left(\frac{\pi^2 E_m I_n}{h^2}\right)\left(1 - 0.577\left(\frac{e}{r}\right)\right)^3 \tag{68.19}$$

$$F_a = \tfrac{1}{4}f_m'\left(1 - \left(\frac{h}{140r}\right)^2\right) \quad [h/r \leq 99] \tag{68.20}$$

$$F_a = \tfrac{1}{4}f_m'\left(\frac{70r}{h}\right)^2 \quad [\text{with } h/r > 99] \tag{68.21}$$

$$F_b = \tfrac{1}{3}f_m' \tag{68.22}$$

10. SD WALL DESIGN: AXIAL COMPRESSION AND FLEXURE—UNREINFORCED

$$P_n = 0.80\left(0.80A_n f_m'\left(1 - \left(\frac{h}{140r}\right)^2\right)\right)$$
$$[\text{for } h/r \leq 99]$$

$$P_n = 0.80\left(0.80A_n f_m'\left(\frac{70r}{h}\right)^2\right)$$
$$[\text{for } h/r > 99]$$

11. ASD WALL DESIGN: AXIAL COMPRESSION AND FLEXURE—REINFORCED

$$F_b = \tfrac{1}{3}f_m' \tag{68.23}$$

$$P_a = (0.25f_m'A_n + 0.65A_{\text{st}}F_s)\left(1 - \left(\frac{h}{140r}\right)^2\right)$$
$$[h/r \leq 99] \tag{68.24}$$

$$P_a = (0.25f_m'A_n + 0.65A_{\text{st}}F_s)\left(\frac{70r}{h}\right)^2$$
$$[h/r > 99] \tag{68.25}$$

$$k_b = \frac{F_b}{F_b + \dfrac{F_s}{n}} \tag{68.26}$$

12. SD WALL DESIGN: AXIAL COMPRESSION AND FLEXURE— REINFORCED

$$M_u = \frac{w_u h^2}{8} + P_{uf}\frac{e_u}{2} + P_u\delta_u \qquad 68.27$$

$$M_u \le \phi M_n \qquad 68.28$$

$$M_n = (A_s f_y + P_u)\left(d - \frac{a}{2}\right) \qquad 68.29$$

$$a = \frac{A_s f_y + P_u}{0.80 f'_m b} \qquad 68.30$$

$$M_{cr} = S f_r \qquad 68.31$$

$$\delta_s = \frac{5 M_{ser} h^2}{48 E_m I_g} \qquad 68.32$$

14. ASD WALL DESIGN: SHEAR WALLS WITH NO NET TENSION— UNREINFORCED

$$f_v = \frac{VQ}{I_n b} = \frac{3V}{2A_n} \quad \text{[rectangular sections]} \qquad 68.34$$

F_v is the least of

(a) $1.5\sqrt{f'_m}$ 68.35

(b) $120 \text{ lbf/in}^2 \ (0.83 \text{ MPa})$ 68.36

(c) $v + 0.45 N_v/A_n$ 68.37

15. SD WALL DESIGN: SHEAR WALLS— UNREINFORCED

(a) $3.8 A_n \sqrt{f'_m}$ 68.38

(b) $300 A_n$ 68.39

(c) $56 A_n + 0.45 N_u \quad \begin{bmatrix} \text{for running bond masonry} \\ \text{not grouted solid} \end{bmatrix}$ 68.40

(d) $90 A_n + 0.45 N_u \quad \begin{bmatrix} \text{for running bond masonry} \\ \text{grouted solid} \end{bmatrix}$ 68.41

16. ASD WALL DESIGN: SHEAR WALLS WITH NET TENSION—REINFORCED

$$f_v = \frac{V}{bd} \qquad 68.42$$

$$F_v = \frac{1}{3}\left(4 - \frac{M}{Vd}\right)\sqrt{f'_m}$$
$$< 80 - 45\left(\frac{M}{Vd}\right) \quad [M/Vd < 1] \qquad 68.43$$

$$F_v = \sqrt{f'_m}$$
$$< 35 \text{ lbf/in}^2 \quad (0.24 \text{ MPa}) \quad [M/Vd \ge 1] \quad 68.44$$

$$F_v = \frac{1}{2}\left(4 - \frac{M}{Vd}\right)\sqrt{f'_m}$$
$$< 120 - 45\left(\frac{M}{Vd}\right) \quad [M/Vd < 1] \qquad 68.45$$

$$F_v = 1.5\sqrt{f'_m}$$
$$< 75 \text{ lbf/in}^2 \quad (0.52 \text{ MPa}) \quad [M/Vd \ge 1] \quad 68.46$$

$$A_v = \frac{Vs}{F_s d} \qquad 68.47$$

17. SD WALL DESIGN: SHEAR WALLS— REINFORCED

$$V_n = V_m + V_s \quad [3.2.4.1.2] \qquad 68.48$$

$$\le 6 A_n \sqrt{f'_m} \quad [\text{when } M_u/V_u d_v < 0.25] \qquad 68.49$$

$$\le 4 A_n \sqrt{f'_m} \quad [\text{when } M_u/V_u d_v > 1.0] \qquad 68.50$$

$$V_m = \left(4 - 1.75\frac{M_u}{V_u d_v}\right) A_n \sqrt{f'_m} + 0.25 P_u \qquad 68.51$$

$$V_m = 2.25 A_n \sqrt{f'_m} \qquad 68.52$$

$$V_s = 0.5\frac{A_v}{s} f_y d_v \qquad 68.53$$

CERM Chapter 69
Masonry Columns

> Chapter, section, equation, figure, and table numbers correspond to CERM. For additional study material, go to the corresponding chapter and section number in CERM.

5. ASD DESIGN FOR PURE COMPRESSION

$$P_a = (0.25 f'_m A_n + 0.65 A_{st} F_s)\left(1 - \left(\frac{h}{140r}\right)^2\right)$$
$$[h/r \le 99] \qquad 69.1$$

$$P_a = (0.25 f'_m A_n + 0.65 A_{st} F_s)\left(\frac{70r}{h}\right)^2$$
$$[h/r > 99] \qquad 69.2$$

6. SD DESIGN FOR PURE COMPRESSION

$$P_n = 0.80 \left(0.80 f'_m (A_n - A_s) + f_y A_s\right)$$
$$\times \left(1 - \left(\frac{h}{140r}\right)\right) \quad \text{[when } h/r \leq 99\text{]} \quad \textit{69.3}$$

$$P_n = 0.80 \left(0.80 f'_m (A_n - A_s) + f_y A_s\right)$$
$$\times \left(\frac{70r}{h}\right)^2 \quad \text{[when } h/r > 99\text{]} \quad \textit{69.4}$$

$$P_u \leq \phi P_n \qquad \textit{69.5}$$

9. BIAXIAL BENDING

$$\frac{1}{P_{\text{biaxial}}} = \frac{1}{P_x} + \frac{1}{P_y} - \frac{1}{P_o} \qquad \textit{69.6}$$

Transportation

CERM Chapter 70
Properties of Solid Bodies

Chapter, section, equation, figure, and table numbers correspond to CERM. For additional study material, go to the corresponding chapter and section number in CERM.

1. CENTER OF GRAVITY

$$x_c = \frac{\int x\, dm}{m} \qquad 70.1$$

$$y_c = \frac{\int y\, dm}{m} \qquad 70.2$$

$$z_c = \frac{\int z\, dm}{m} \qquad 70.3$$

$$x_c = \frac{\sum m_i x_{ci}}{\sum m_i} \qquad 70.4$$

$$y_c = \frac{\sum m_i y_{ci}}{\sum m_i} \qquad 70.5$$

$$z_c = \frac{\sum m_i z_{ci}}{\sum m_i} \qquad 70.6$$

2. MASS AND WEIGHT

$$m = \rho V \qquad 70.7$$

$$w = \frac{mg}{g_c} \qquad 70.8(b)$$

4. MASS MOMENT OF INERTIA

$$I_x = \int (y^2 + z^2)\, dm \qquad 70.9$$

$$I_y = \int (x^2 + z^2)\, dm \qquad 70.10$$

$$I_z = \int (x^2 + y^2)\, dm \qquad 70.11$$

5. PARALLEL AXIS THEOREM

$$I_{\text{any parallel axis}} = I_c + md^2 \qquad 70.12$$

$$I = I_{c,1} + m_1 d_1^2 + I_{c,2} + m_2 d_2^2 + \cdots \qquad 70.13$$

6. RADIUS OF GYRATION

$$k = \sqrt{\frac{I}{m}} \qquad 70.14$$

$$I = k^2 m \qquad 70.15$$

CERM Chapter 71
Kinematics

Chapter, section, equation, figure, and table numbers correspond to CERM. For additional study material, go to the corresponding chapter and section number in CERM.

5. LINEAR PARTICLE MOTION

$$s(t) = \int v(t)\, dt = \int \left(\int a(t)\, dt \right) dt \qquad 71.2$$

$$v(t) = \frac{ds(t)}{dt} = \int a(t)\, dt \qquad 71.3$$

$$a(t) = \frac{dv(t)}{dt} = \frac{d^2 s(t)}{dt^2} \qquad 71.4$$

$$v_{\text{ave}} = \frac{\int_1^2 v(t)\, dt}{t_2 - t_1} = \frac{s_2 - s_1}{t_2 - t_1} \qquad 71.5$$

$$a_{\text{ave}} = \frac{\int_1^2 a(t)\, dt}{t_2 - t_1} = \frac{v_2 - v_1}{t_2 - t_1} \qquad 71.6$$

6. DISTANCE AND SPEED

$$\text{displacement} = s(t_2) - s(t_1) \qquad 71.7$$

7. UNIFORM MOTION

$$s(t) = s_0 + \mathrm{v}t \qquad 71.8$$

$$\mathrm{v}(t) = \mathrm{v} \qquad 71.9$$

$$a(t) = 0 \qquad 71.10$$

8. UNIFORM ACCELERATION

$$a(t) = a \qquad 71.11$$

$$\mathrm{v}(t) = a \int dt = \mathrm{v}_0 + at \qquad 71.12$$

$$s(t) = a \iint dt^2 = s_0 + \mathrm{v}_0 t + \tfrac{1}{2}at^2 \qquad 71.13$$

Table 71.1 Uniform Acceleration Formulas[a]

to find	given these	use this equation
a	$t, \mathrm{v}_0, \mathrm{v}$	$a = \dfrac{\mathrm{v} - \mathrm{v}_0}{t}$
a	t, v_0, s	$a = \dfrac{2s - 2\mathrm{v}_0 t}{t^2}$
a	$\mathrm{v}_0, \mathrm{v}, s$	$a = \dfrac{\mathrm{v}^2 - \mathrm{v}_0^2}{2s}$
s	t, a, v_0	$s = \mathrm{v}_0 t + \tfrac{1}{2}at^2$
s	$a, \mathrm{v}_0, \mathrm{v}$	$s = \dfrac{\mathrm{v}^2 - \mathrm{v}_0^2}{2a}$
s	$t, \mathrm{v}_0, \mathrm{v}$	$s = \tfrac{1}{2}t(\mathrm{v}_0 + \mathrm{v})$
t	$a, \mathrm{v}_0, \mathrm{v}$	$t = \dfrac{\mathrm{v} - \mathrm{v}_0}{a}$
t	a, v_0, s	$t = \dfrac{\sqrt{\mathrm{v}_0^2 + 2as} - \mathrm{v}_0}{a}$
t	$\mathrm{v}_0, \mathrm{v}, s$	$t = \dfrac{2s}{\mathrm{v}_0 + \mathrm{v}}$
v_0	t, a, v	$\mathrm{v}_0 = \mathrm{v} - at$
v_0	t, a, s	$\mathrm{v}_0 = \dfrac{s}{t} - \tfrac{1}{2}at$
v_0	a, v, s	$\mathrm{v}_0 = \sqrt{\mathrm{v}^2 - 2as}$
v	t, a, v_0	$\mathrm{v} = \mathrm{v}_0 + at$
v	a, v_0, s	$\mathrm{v} = \sqrt{\mathrm{v}_0^2 + 2as}$

[a]The table can be used for rotational problems by substituting α, ω, and θ for a, v, and s, respectively.

CERM Chapter 72
Kinetics

> Chapter, section, equation, figure, and table numbers correspond to CERM. For additional study material, go to the corresponding chapter and section number in CERM.

5. LINEAR MOMENTUM

$$\mathbf{p} = \frac{m\mathbf{v}}{g_c} \qquad 72.1(b)$$

9. NEWTON'S SECOND LAW OF MOTION

$$F = \left(\frac{m}{g_c}\right)\left(\frac{dv}{dt}\right) = \frac{ma}{g_c} \qquad 72.12(b)$$

$$T = \left(\frac{I}{g_c}\right)\left(\frac{d\omega}{dt}\right) = \frac{I\alpha}{g_c} \qquad 72.14(b)$$

10. CENTRIPETAL FORCE

$$F_c = \frac{m\mathrm{v}_t^2}{g_c r} \qquad 72.15(b)$$

13. FLAT FRICTION

$$N = \frac{mg}{g_c} \qquad 72.19(b)$$

$$N = \frac{mg\cos\phi}{g_c} \qquad 72.20(b)$$

$$F_{f,\mathrm{max}} = f_s N \qquad 72.21$$

$$\tan\phi = f_s \qquad 72.22$$

16. ROLLING RESISTANCE

$$F_r = \frac{mga}{rg_c} = \frac{wa}{r} \qquad 72.27(b)$$

$$f_r = \frac{F_r}{w} = \frac{a}{r} \qquad 72.28$$

17. MOTION OF RIGID BODIES

$$\sum F_y = ma_y \quad \text{[consistent units]} \qquad 72.30$$

$$T = \frac{I\alpha}{g_c} \qquad 72.31(b)$$

20. IMPULSE

$$\text{Imp} = \int_{t_1}^{t_2} F \, dt \quad \text{[linear]} \qquad 72.44$$

$$\text{Imp} = \int_{t_1}^{t_2} T \, dt \quad \text{[angular]} \qquad 72.45$$

$$\text{Imp} = F(t_2 - t_1) \quad \text{[linear]} \qquad 72.46$$

$$\text{Imp} = T(t_2 - t_1) \quad \text{[angular]} \qquad 72.47$$

21. IMPULSE-MOMENTUM PRINCIPLE

$$F(t_2 - t_1) = \frac{m(\text{v}_2 - \text{v}_1)}{g_c} \qquad 72.50(b)$$

$$T(t_2 - t_1) = \frac{I(\omega_2 - \omega_1)}{g_c} \qquad 72.51(b)$$

22. IMPULSE-MOMENTUM PRINCIPLE IN OPEN SYSTEMS

$$F = \frac{m\Delta\text{v}}{g_c \Delta t} = \frac{\dot{m}\Delta\text{v}}{g_c} \qquad 72.52(b)$$

23. IMPACTS

$$m_1\text{v}_1 + m_2\text{v}_2 = m_1\text{v}_1' + m_2\text{v}_2' \qquad 72.53$$

Figure 72.17 *Direct Central Impact*

$$m_1\text{v}_1^2 + m_2\text{v}_2^2 = m_1\text{v}_1'^2 + m_2\text{v}_2'^2 \Big|_{\text{elastic impact}} \qquad 72.54$$

24. COEFFICIENT OF RESTITUTION

$$e = \frac{\text{relative separation velocity}}{\text{relative approach velocity}}$$

$$= \frac{\text{v}_1' - \text{v}_2'}{\text{v}_2 - \text{v}_1} \qquad 72.55$$

33. ROADWAY BANKING

$$\tan\phi = \frac{\text{v}_t^2}{gr} \quad \text{[no friction]} \qquad 72.90$$

$$\text{v}_t = \sqrt{gr\tan\phi} \qquad 72.91$$

$$e = \tan\phi = \frac{\text{v}_t^2 - fgr}{gr + f\text{v}_t^2} \quad \text{[with friction]} \qquad 72.92$$

Figure 72.24 *Roadway Banking*

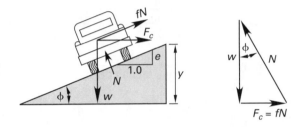

CERM Chapter 73
Roads and Highways: Capacity Analysis

Chapter, section, equation, figure, and table numbers correspond to CERM. For additional study material, go to the corresponding chapter and section number in CERM.

8. VOLUME PARAMETERS

$$K = \frac{\text{DHV}}{\text{AADT}} \qquad 73.1$$

$$\text{DDHV} = D(\text{DHV}) = DK(\text{AADT}) \qquad 73.2$$

$$\text{volume-capacity ratio}_i = (v/c)_i \qquad 73.3$$

$$v_{m,i} = c\,(v/c)_{m,i} \qquad 73.4$$

$$\text{PHF} = \frac{\text{actual hourly volume}_\text{vph}}{\text{peak rate of flow}_\text{vph}}$$

$$= \frac{V_\text{vph}}{v_p} = \frac{V_\text{vph}}{4V_{15\,\text{min,peak}}} \qquad 73.5$$

$$V_i = (v_{m,i})N(\text{adjustment factors}) \qquad 73.6$$

9. TRIP GENERATION

$$\text{no. of trips} = a + b\left(\frac{\text{calling population}}{\text{parameter}}\right) \qquad 73.7$$

$$\log(\text{no. of trips}) = A + B\log\left(\frac{\text{calling population}}{\text{parameter}}\right) \qquad 73.8(a)$$

$$\text{no. of trips} = \frac{C}{\left(\dfrac{\text{calling population}}{\text{parameter}}\right)^D} \qquad 73.8(b)$$

10. SPEED, FLOW, AND DENSITY RELATIONSHIPS

$$S = S_f \left(1 - \frac{D}{D_j} \right) \qquad 73.9$$

$$v = SD = \frac{3600 \frac{\text{sec}}{\text{hr}}}{\text{headway} \frac{\text{sec}}{\text{veh}}} \qquad 73.10$$

$$\text{spacing}_{\text{ft/veh}} = \frac{5280 \frac{\text{ft}}{\text{mi}}}{D_{\text{vpm/lane}}} \qquad 73.11(b)$$

$$\text{headway}_{\text{sec/veh}} = \frac{\text{spacing}_{\text{ft/veh}}}{\text{space mean speed}_{\text{ft/sec}}} \qquad 73.12(b)$$

$$v_{\text{vph}} = \frac{3600 \frac{\text{sec}}{\text{hr}}}{\text{headway}_{\text{sec/veh}}} \qquad 73.13$$

13. FREEWAYS

$$D = \frac{v_p}{S} \qquad 73.14$$

$$v = c \left(\frac{v}{c} \right) \qquad 73.15$$

$$v_{m,i} = c \left(\frac{v}{c} \right)_{m,i} \qquad 73.16$$

$$V = v_p(\text{PHF})N f_{\text{HV}} f_p \qquad 73.17$$

$$f_{\text{HV}} = \frac{1}{1 + P_T(E_T - 1) + P_R(E_R - 1)} \qquad 73.18$$

$$\text{FFS} = \text{BFFS} - f_{\text{LW}} - f_{\text{LC}} - f_N - f_{\text{ID}} \qquad 73.19$$

14. MULTILANE HIGHWAYS

$$\text{FFS} = \text{BFFS} - f_M - f_{\text{LW}} - f_{\text{LC}} - f_A \qquad 73.20$$

$$v_p = D_m(\text{FFS}) = \frac{V}{N(\text{PHF})f_{\text{HV}} f_p} \qquad 73.21$$

15. SIGNALIZED INTERSECTIONS

$$c_i = s_i \left(\frac{g_i}{C} \right) \qquad 73.22$$

$$s_{\text{vphgpl}} = \frac{3600 \frac{\text{sec}}{\text{hr}}}{\text{saturation headway}_{\text{sec/veh}}} \qquad 73.23$$

$$X_i = \left(\frac{v}{c} \right)_i = \frac{v_i}{s_i \left(\frac{g_i}{C} \right)} = \frac{v_i C}{s_i g_i} \qquad 73.24$$

$$X_c = \sum_i \left(\frac{v}{s} \right)_{ci} \left(\frac{C}{C - L} \right) \qquad 73.25$$

$$s = s_o N f_w f_{\text{HV}} f_g f_p f_{\text{bb}} f_a f_{\text{LU}} f_{\text{RT}} f_{\text{LT}} f_{\text{Lpb}} f_{\text{Rpb}} \qquad 73.26$$

19. TRAFFIC-ACTIVATED TIMING

$$\text{no. of cars in initial period} = \frac{\text{distance between line and detector}}{\text{car length}} \qquad 73.28$$

20. WALKWAYS

$$v = \frac{v_{p,15}}{15W_E} \qquad 73.29$$

$$v = SD = \frac{S}{M} \qquad 73.30$$

28. QUEUING MODELS

$$L = \lambda W \qquad 73.31$$

$$L_q = \lambda W_q \qquad 73.32$$

$$W = W_q + \frac{1}{\mu} \qquad 73.33$$

29. M/M/1 SINGLE-SERVER MODEL

$$f(t) = \mu e^{-\mu t} \qquad 73.34$$

$$P\{t > h\} = e^{-\mu h} \qquad 73.35$$

$$p\{x\} = \frac{e^{-\lambda} \lambda^x}{x!} \qquad 73.36$$

$$p\{0\} = 1 - \rho \qquad 73.37$$

$$p\{n\} = p\{0\}\rho^n \qquad 73.38$$

$$W = \frac{1}{\mu - \lambda}$$
$$= W_q + \frac{1}{\mu} = \frac{L}{\lambda} \qquad 73.39$$

$$W_q = \frac{\rho}{\mu - \lambda} = \frac{L_q}{\lambda} \qquad 73.40$$

$$L = \frac{\lambda}{\mu - \lambda} = L_q + \rho \qquad 73.41$$

$$L_q = \frac{\rho \lambda}{\mu - \lambda} \qquad 73.42$$

30. M/M/s MULTI-SERVER MODEL

$$W = W_q + \frac{1}{\mu} \qquad 73.43$$

$$W_q = \frac{L_q}{\lambda} \qquad 73.44$$

$$L_q = \frac{p\{0\}\rho\left(\frac{\lambda}{\mu}\right)^s}{s!(1-\rho)^2} \qquad 73.45$$

$$L = L_q + \frac{\lambda}{\mu} \qquad 73.46$$

$$p\{0\} = \frac{1}{\dfrac{\left(\frac{\lambda}{\mu}\right)^s}{s!\left(1-\frac{\lambda}{s\mu}\right)} + \displaystyle\sum_{j=0}^{s-1}\dfrac{\left(\frac{\lambda}{\mu}\right)^j}{j!}} \qquad 73.47$$

$$p\{n\} = \frac{p\{0\}\left(\frac{\lambda}{\mu}\right)^n}{n!} \quad [n \le s] \qquad 73.48$$

$$p\{n\} = \frac{p\{0\}\left(\frac{\lambda}{\mu}\right)^n}{s!s^{n-s}} \quad [n > s] \qquad 73.49$$

33. TEMPORARY TRAFFIC CONTROL ZONES

$$L_{\text{ft}} = \frac{W_{\text{ft}}S_{\text{mph}}^2}{60} \quad [S \le 40 \text{ mph}]$$
$$73.50(b)$$

$$L_{\text{ft}} = W_{\text{ft}}S_{\text{mph}} \quad [S > 45 \text{ mph}]$$
$$73.51(b)$$

CERM Chapter 74
Vehicle Dynamics and Accident Analysis

Chapter, section, equation, figure, and table numbers correspond to CERM. For additional study material, go to the corresponding chapter and section number in CERM.

1. VEHICLE DYNAMICS

$$F_i = \frac{ma}{g_c} = \frac{wa}{g} \qquad 74.1(b)$$

$$F_g = w \sin \phi$$
$$\approx w \tan \phi = \frac{wG\%}{100} \qquad 74.2$$

$$F_r = f_r w \cos \phi$$
$$\approx f_r w \qquad 74.3$$

$$F_D = \frac{C_D A \rho v^2}{2g_c} \qquad 74.4(b)$$

$$F_D = K A v^2 \approx 0.0006 A_{\text{ft}^2} v_{\text{mi/hr}}^2$$
$$74.5(b)$$

$$P = (F_i + F_g + F_r + F_c + F_D)v \qquad 74.6$$

$$v_{\text{vehicle}} = v_{t,\text{tire}} = \pi d_{\text{tire}} n_{\text{tire}} \qquad 74.7$$

$$n_{\text{wheel}} = \frac{n_{\text{engine}}}{R_{\text{transmission}}R_{\text{differential}}} \qquad 74.8$$

$$R_{\text{transmission}} = \frac{n_{\text{engine}}}{n_{\text{driveshaft}}} \qquad 74.9$$

$$R_{\text{differential}} = \frac{n_{\text{driveshaft}}}{n_{\text{wheel}}} \qquad 74.10$$

$$P_{\text{hp}} = \frac{T_{\text{in-lbf}}n_{\text{rpm}}}{63{,}025} \qquad 74.11(b)$$

$$F_{\text{tractive}} = \frac{\eta_m T R_{\text{transmission}}R_{\text{differential}}}{r_{\text{tire}}} \qquad 74.12$$

$$v = \frac{2\pi n_{\text{rev/sec}}(1-i)}{R_{\text{transmission}}R_{\text{differential}}} \qquad 74.13$$

$$\dot{m}_{\text{fuel,lbm/hr}} = P_{\text{brake,hp}}(\text{BSFC}_{\text{lbm/hp-hr}})$$
$$74.14(b)$$

$$\dot{Q}_{\text{fuel,gal/hr}} = \frac{\dot{m}_{\text{fuel,lbm/hr}}\left(7.48\,\dfrac{\text{gal}}{\text{ft}^3}\right)}{\rho_{\text{lbm/ft}^3}} \qquad 74.15(b)$$

$$s'_{\text{fuel,mi/gal}} = \frac{v_{\text{mi/hr}}}{\dot{Q}_{\text{fuel,gal/hr}}} \qquad 74.16(b)$$

2. DYNAMICS OF STEEL-WHEELED RAILROAD ROLLING STOCK

$$\text{DBP} = \text{TFDA} - \text{LR}$$
$$= \text{CR} + \text{AF} \qquad 74.17$$

$$F_{\text{tractive,lbf}} = \frac{375 P_{\text{hp,rated}}\eta_{\text{drive}}}{v_{\text{mph}}} \qquad 74.18(b)$$

$$R_{\text{lbf/ton}} = 0.6 + \frac{20}{w_{\text{tons}}} + 0.01 v_{\text{mph}}$$
$$+ \frac{K v_{\text{mph}}^2}{w_{\text{tons}}n} \qquad 74.19$$

3. COEFFICIENT OF FRICTION

$$f = \frac{F_f}{N} \qquad \text{74.20}$$

5. STOPPING DISTANCE

$$s_{\text{stopping}} = vt_p + s_b \qquad \text{74.21}$$

6. BRAKING AND DECELERATION RATE

$$a = fg = f\left(32.2 \ \frac{\text{ft}}{\text{sec}^2}\right) \qquad \text{74.22(b)}$$

7. BRAKING AND SKIDDING DISTANCE

$$s_b = \frac{v^2}{2g(f\cos\theta + \sin\theta)} \qquad \text{74.23}$$

$$s_b = \frac{v^2}{2g(f+G)}$$
$$\approx \frac{v_{\text{mph}}^2}{30(f+G)} \qquad \text{74.24(b)}$$

9. ANALYSIS OF ACCIDENT DATA

$$R = \frac{(\text{no. of accidents})(10^6)}{(\text{ADT})(\text{no. of years})\left(365 \ \dfrac{\text{days}}{\text{yr}}\right)} \qquad \text{74.25}$$

$$R = \frac{(\text{no. of accidents})(10^8)}{(\text{ADT})(\text{no. of years})\left(365 \ \dfrac{\text{days}}{\text{yr}}\right)L_{\text{mi}}} \qquad \text{74.26}$$

CERM Chapter 75
Flexible Pavement Design

Chapter, section, equation, figure, and table numbers correspond to CERM. For additional study material, go to the corresponding chapter and section number in CERM.

9. WEIGHT-VOLUME RELATIONSHIPS

$$P_b = (100\%)(\text{aggregate surface area})$$
$$\times (\text{asphalt thickness})$$
$$\times (\text{specific weight of asphalt}) \qquad \text{75.1}$$

10. PLACEMENT AND PAVING EQUIPMENT

$$L_{\text{ft/ton}} = \frac{2000 \ \dfrac{\text{lbf}}{\text{ton}}}{w_{\text{ft}}t_{\text{ft}}\gamma_{\text{compacted,lbf/ft}^3}} \qquad \text{75.2(b)}$$

$$L_{\text{ft/ton}} = \frac{18,000 \ \dfrac{\text{lbf-ft}^2}{\text{ton-yd}^2}}{r_{\text{lbf/yd}^2}w_{\text{ft}}} \qquad \text{75.3(b)}$$

$$v_{\text{ft/min}} = \frac{R_{p,\text{ton/hr}}L_{\text{ft/ton}}}{60 \ \dfrac{\text{min}}{\text{hr}}}$$
$$= \frac{R_{p,\text{ton/hr}}\left(2000 \ \dfrac{\text{lbf}}{\text{ton}}\right)\left(12 \ \dfrac{\text{in}}{\text{ft}}\right)}{w_{\text{ft}}t_{\text{in}}\gamma_{\text{lbf/ft}^3}\left(60 \ \dfrac{\text{min}}{\text{hr}}\right)} \qquad \text{75.4(b)}$$

12. CHARACTERISTICS OF ASPHALT CONCRETE

$$G_{\text{sa}} = \frac{W}{V_{\text{aggregate}}\gamma_{\text{water}}} \qquad \text{75.5(b)}$$

$$G_{\text{sb}} = \frac{P_1 + P_2 + \dots P_n}{\dfrac{P_1}{G_1} + \dfrac{P_2}{G_2} + \dots + \dfrac{P_n}{G_n}} \qquad \text{75.6}$$

$$G_{\text{sa}} = \frac{A}{A-C} \qquad \text{75.7}$$

$$G_{\text{sb}} = \frac{A}{B-C} \qquad \text{75.8}$$

$$\text{absorption} = \frac{(100\%)(B-A)}{A} \qquad \text{75.9}$$

$$G_{\text{sa}} = \frac{A}{B+A-C} \qquad \text{75.10}$$

$$G_{\text{sb}} = \frac{A}{B+S-C} \qquad \text{75.11}$$

$$\text{absorption} = \frac{(100\%)(S-A)}{A} \qquad \text{75.12}$$

$$G_{\text{se}} = \frac{100\% - P_b}{\dfrac{100\%}{G_{\text{mm}}} - \dfrac{P_b}{G_b}} \qquad \text{75.13}$$

$$G_{\text{mm}} = \frac{100\%}{\dfrac{P_s}{G_{\text{se}}} + \dfrac{P_b}{G_b}} \qquad \text{75.14}$$

$$G_{\text{mm}} = \frac{A}{A+D-E} \qquad \text{75.15}$$

$$P_{\text{ba}} = \frac{(100\%)G_b(G_{\text{se}} - G_{\text{sb}})}{G_{\text{sb}}G_{\text{se}}} \qquad \text{75.16}$$

$$P_{\text{be}} = P_b - \frac{P_{\text{ba}}P_s}{100\%} \qquad 75.17$$

$$\text{VMA} = 100\% - \frac{G_{\text{mb}}P_s}{G_{\text{sb}}} \qquad 75.18$$

$$P_a = \text{VTM} = \frac{(100\%)(G_{\text{mm}} - G_{\text{mb}})}{G_{\text{mm}}} \quad 75.19$$

$$\text{VFA} = \frac{(100\%)(\text{VMA} - \text{VTM})}{\text{VMA}} \qquad 75.20$$

13. MARSHALL MIX TEST PROCEDURE

Figure 75.2 Volumes in a Compacted Asphalt Specimen

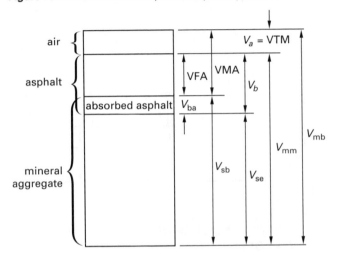

VMA = volume of voids in mineral aggregate
V_{mb} = bulk volume of compacted mix
V_{mm} = voidless volume of paving mix
V_a = volume of air voids
V_b = volume of asphalt
V_{ba} = volume of absorbed asphalt
V_{sb} = volume of mineral aggregate (by bulk specific gravity)
V_{se} = volume of mineral aggregate (by effective specific gravity)

15. TRUCK FACTORS

$$\text{TF} = \frac{\text{ESALs}}{\text{no. of trucks}} \qquad 75.22$$

16. DESIGN TRAFFIC

$$\text{ESAL}_{20} = (\text{ESAL}_{\text{first year}})(\text{GF}) \qquad 75.23$$

$$w_{18} = D_D D_L \hat{w}_{18} \qquad 75.24$$

21. PAVEMENT STRUCTURAL NUMBER

$$\text{SN} = D_1 a_1 + D_2 a_2 m_2 + D_3 a_3 m_3 \qquad 75.29$$

$$a_2 = 0.249(\log_{10} E_{\text{BS}}) - 0.977$$
$$[AASHTO\ Guide\ \text{p. II-20}] \qquad 75.30$$

$$a_3 = 0.227(\log_{10} E_{\text{SB}}) - 0.839$$
$$[AASHTO\ Guide\ \text{p. II-22}] \qquad 75.31$$

CERM Chapter 76
Rigid Pavement Design

> Chapter, section, equation, figure, and table numbers correspond to CERM. For additional study material, go to the corresponding chapter and section number in CERM.

4. LAYER MATERIAL STRENGTHS

$$k = \frac{M_R}{19.4} \qquad 76.1$$

$$E_c = 57,000\sqrt{f_c'} \quad [E_c\ \text{and}\ f_c'\ \text{in lbf/in}^2] \qquad 76.2$$

8. STEEL REINFORCING

$$P_s = \frac{L_{\text{ft}}F(100\%)}{2f_{s,\text{lbf/in}^2}} \qquad 76.3$$

$$P_t = \frac{A_s(100\%)}{YD} \qquad 76.4$$

CERM Chapter 77
Plane Surveying

> Chapter, section, equation, figure, and table numbers correspond to CERM. For additional study material, go to the corresponding chapter and section number in CERM.

1. ERROR ANALYSIS: MEASUREMENTS OF EQUAL WEIGHT

$$x_p = \frac{x_1 + x_2 + \cdots + x_k}{k} \qquad 77.1$$

$$E_{\text{mean}} = \frac{0.6745s}{\sqrt{k}}$$
$$= \frac{E_{\text{total},k\ \text{measurements}}}{\sqrt{k}} \qquad 77.2$$

2. ERROR ANALYSIS: MEASUREMENTS OF UNEQUAL WEIGHT

$$E_{p,\text{weighted}} = 0.6745\sqrt{\frac{\sum \left(w_i(\overline{x} - x_i)^2\right)}{(k-1)\sum w_i}} \qquad 77.3$$

3. ERRORS IN COMPUTED QUANTITIES

$$E_{\text{sum}} = \sqrt{E_1^2 + E_2^2 + E_3^2 + \cdots} \qquad \textit{77.4}$$

$$E_{\text{product}} = \sqrt{x_1^2 E_2^2 + x_2^2 E_1^2} \qquad \textit{77.5}$$

13. DISTANCE MEASUREMENT: TAPING

$$C_T = L\alpha(T - T_s) \quad \text{[temperature]} \qquad \textit{77.6}$$

$$C_P = \frac{(P - P_s)L}{AE} \quad \text{[tension]} \qquad \textit{77.7}$$

$$C_s = \pm \left(\frac{W^2 L^3}{24 P^2} \right) \quad \text{[sag]} \qquad \textit{77.8}$$

14. DISTANCE MEASUREMENT: TACHYOMETRY

$$x = K(R_2 - R_1) + C \qquad \textit{77.9}$$

$$x = K(R_2 - R_1)\cos^2 \theta + C \cos \theta \qquad \textit{77.10}$$

$$y = \tfrac{1}{2} K(R_2 - R_1)\sin 2\theta + C \sin \theta \qquad \textit{77.11}$$

Figure 77.1 *Horizontal Stadia Measurement*

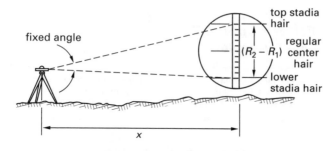

18. ELEVATION MEASUREMENT

$$h_c = \left(2.4 \times 10^{-8} \, \frac{1}{\text{ft}} \right) x_{\text{ft}}^2 \qquad \textit{77.12}$$

$$h_r = \left(3.0 \times 10^{-9} \, \frac{1}{\text{ft}} \right) x_{\text{ft}}^2 \qquad \textit{77.13}$$

$$h_a = R_{\text{observed}} + h_r - h_c$$
$$= R_{\text{observed}} - \left(2.1 \times 10^{-8} \, \frac{1}{\text{ft}} \right) x_{\text{ft}}^2 \qquad \textit{77.14}$$

19. ELEVATION MEASUREMENT: DIRECT LEVELING

$$y_{\text{A-L}} = R_A - h_{\text{rc,A-L}} - \text{HI} \qquad \textit{77.15}$$

$$y_{\text{L-B}} = \text{HI} + h_{\text{rc,L-B}} - R_B \qquad \textit{77.16}$$

$$y_{\text{A-B}} = y_{\text{A-L}} + y_{\text{L-B}}$$
$$= R_A - R_B + h_{\text{rc,L-B}} - h_{\text{rc,A-L}} \qquad \textit{77.17}$$

$$y_{\text{A-B}} = R_A - R_B \qquad \textit{77.18}$$

Figure 77.4 *Direct Leveling*

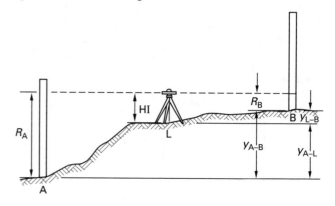

21. ELEVATION MEASUREMENT: INDIRECT LEVELING

$$y_{\text{A-B}} = \text{AC} \tan \alpha + 2.1 \times 10^{-8} (\text{AC})^2 \qquad \textit{77.19}$$

$$y_{\text{A-B}} = \text{AC} \tan \beta - 2.1 \times 10^{-8} (\text{AC})^2 \qquad \textit{77.20}$$

$$y_{\text{A-B}} = \tfrac{1}{2} \text{AC}(\tan \alpha + \tan \beta) \qquad \textit{77.21}$$

Figure 77.5 *Indirect Leveling*

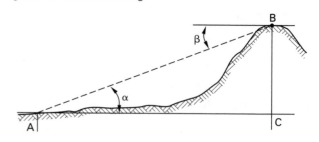

23. DIRECTION SPECIFICATION

Figure 77.6 *Calculation of Bearing Angle*

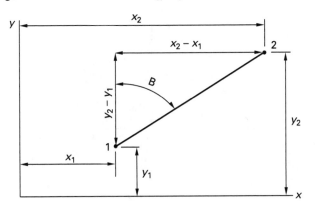

$$\tan B = \frac{x_2 - x_1}{y_2 - y_1} \qquad 77.22$$

$$D = \sqrt{(y_2 - y_1)^2 + (x_2 - x_1)^2} \qquad 77.23$$

$$\tan A_N = \frac{x_2 - x_1}{y_2 - y_1} \qquad 77.24$$

$$\tan A_S = \frac{x_1 - x_2}{y_1 - y_2} \qquad 77.25$$

25. TRAVERSES

1. The sum of the deflection angles is $360°$.
2. The sum of the interior angles of a polygon with n sides is $(n-2)(180°)$.

28. BALANCING CLOSED TRAVERSE DISTANCES

$$L = \sqrt{\begin{array}{c}(\text{closure in departure})^2 \\ + (\text{closure in latitude})^2\end{array}} \qquad 77.28$$

$$\frac{\text{leg departure correction}}{\text{closure in departure}} = \frac{-\text{leg length}}{\text{total traverse length}} \qquad 77.29$$

$$\frac{\text{leg latitude correction}}{\text{closure in latitude}} = \frac{-\text{leg length}}{\text{total traverse length}} \qquad 77.30$$

$$\frac{\text{leg departure correction}}{\text{closure in departure}}$$
$$= -\left(\frac{\text{leg departure}}{\text{sum of departure absolute values}}\right) \qquad 77.31$$

$$\frac{\text{leg latitude correction}}{\text{closure in latitude}}$$
$$= -\left(\frac{\text{leg latitude}}{\text{sum of latitude absolute values}}\right) \qquad 77.32$$

30. TRAVERSE AREA: METHOD OF COORDINATES

$$A = \frac{1}{2} \left| \left(\sum_{i=1}^{n} y_i (x_{i-1} - x_{i+1}) \right) \right| \qquad 77.33$$

$$\frac{x_1}{y_1} \diagdown \frac{x_2}{y_2} \diagdown \frac{x_3}{y_3} \diagdown \frac{x_4}{y_4} \diagdown \frac{x_1}{y_1} \qquad \text{etc.}$$

$$A = \frac{1}{2} \left| \begin{array}{c} \sum \text{ of full line products} \\ - \sum \text{ of broken line products} \end{array} \right| \qquad 77.34$$

31. TRAVERSE AREA: DOUBLE MERIDIAN DISTANCE

$$\text{DMD}_{\text{leg } i} = \text{DMD}_{\text{leg } i-1} + \text{departure}_{\text{leg } i-1}$$
$$+ \text{departure}_{\text{leg } i} \qquad 77.35$$

$$A = \frac{1}{2} \left| \sum (\text{latitude}_{\text{leg } i} \times \text{DMD}_{\text{leg } i}) \right| \qquad 77.36$$

32. AREAS BOUNDED BY IRREGULAR BOUNDARIES

$$A = d \left(\frac{h_1 + h_n}{2} + \sum_{i=2}^{n-1} h_i \right) \qquad 77.37$$

$$A = \left(\frac{d}{3}\right) h_1 - h_n + 2 \underset{\substack{\text{for all odd numbers} \\ \text{starting at } i=3}}{\sum_{i=3}^{n-1} h_i} + 4 \underset{\substack{\text{for all even numbers} \\ \text{starting at } i=2}}{\sum_{i=2}^{n-1} h_i}$$
$$77.38$$

33. PHOTOGRAMMETRY

$$\text{scale} = \frac{\text{focal length}}{\text{flight altitude}}$$
$$= \frac{\text{length in photograph}}{\text{true length}} \qquad 77.39$$

Relief Displacement *(bonus material)*

$$\text{actual height} = \frac{\left(\begin{array}{c}\text{observed image displacement,} \\ \text{base to top of object}\end{array}\right) \times \left(\begin{array}{c}\text{camera altitude} \\ \text{above base of object}\end{array}\right)}{\begin{array}{c}\text{radial distance,} \\ \text{nadir to top of object}\end{array}}$$

Parallax Imagery (bonus material)

$$\text{actual height} = \dfrac{\left(\begin{array}{c}\text{differential}\\\text{parallax}\end{array}\right) \times \left(\begin{array}{c}\text{camera altitude}\\\text{above base of object}\end{array}\right)}{\begin{array}{c}\text{stereoscopic parallax}\\\text{at base of object}\\ + \text{ differential parallax}\end{array}}$$

CERM Chapter 78
Horizontal, Compound, Vertical, and Spiral Curves

Chapter, section, equation, figure, and table numbers correspond to CERM. For additional study material, go to the corresponding chapter and section number in CERM.

1. HORIZONTAL CURVES

$$R = \frac{5729.578}{D} \quad \text{[arc definition]} \qquad 78.1$$

$$R = \frac{50}{\sin\dfrac{D}{2}} \quad \text{[chord definition]} \qquad 78.2$$

$$L = \frac{2\pi R I}{360°} = R I_{\text{radians}} = \frac{100 I}{D} \qquad 78.3$$

$$T = R\tan\frac{I}{2} \qquad 78.4$$

$$E = R\left(\sec\frac{I}{2} - 1\right) = R\tan\frac{I}{2}\tan\frac{I}{4} \qquad 78.5$$

$$\text{HSO} = R\left(1 - \cos\frac{I}{2}\right) = \frac{C}{2}\tan\frac{I}{4} \qquad 78.6$$

$$C = 2R\sin\frac{I}{2} = 2T\cos\frac{I}{2} \qquad 78.7$$

Figure 78.1 Horizontal Curve Elements

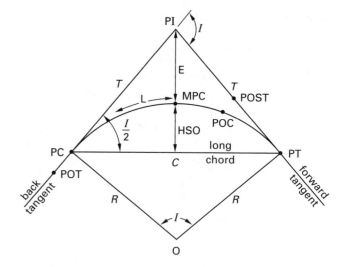

2. DEGREE OF CURVE

$$D = \frac{(360°)(100)}{2\pi R} = \frac{5729.578}{R} \quad \text{[arc basis]} \qquad 78.8$$

$$\sin\frac{D}{2} = \frac{50}{R} \quad \text{[chord basis]} \qquad 78.9$$

$$L \approx \left(\frac{I}{D}\right)(100 \text{ ft}) \qquad 78.10$$

3. STATIONING ON A HORIZONTAL CURVE

$$\text{sta PT} = \text{sta PC} + L \qquad 78.11$$
$$\text{sta PC} = \text{sta PI} - T \qquad 78.12$$

4. CURVE LAYOUT BY DEFLECTION ANGLE

1. The deflection angle between a tangent and a chord (Fig. 78.3(a)) is half of the arc's subtended angle.

2. The angle between two chords (Fig. 78.3(b)) is half of the arc's subtended angle.

$$\alpha = \angle\text{V-PC-A} = \frac{\beta}{2} \qquad 78.13$$

$$\frac{\beta}{360°} = \frac{\text{arc length PC-A}}{2\pi R} \qquad 78.14$$

$$\frac{\beta}{I} = \frac{\text{arc length PC-A}}{L} \qquad 78.15$$

$$C_{\text{PC-A}} = 2R\sin\alpha = 2R\sin\frac{\beta}{2} \qquad 78.16$$

Figure 78.3 *Circular Curve Deflection Angle*

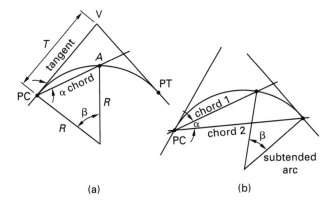

(a)   (b)

5. TANGENT OFFSETS

$$y = R(1 - \cos \beta)$$
$$= R - \sqrt{R^2 - x^2} \qquad \text{78.17}$$

$$\beta = \arcsin \frac{x}{R} = \arccos \left(\frac{R - y}{R} \right) \qquad \text{78.18}$$

$$x = R \sin \beta = \sqrt{2Ry - y^2} \qquad \text{78.19}$$

Figure 78.4 *Tangent Offset*

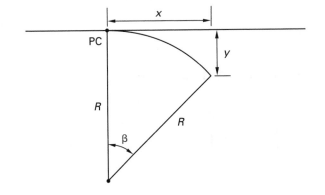

6. CURVE LAYOUT BY TANGENT OFFSETS

Figure 78.5 *Tangent and Chord Offset Geometry*

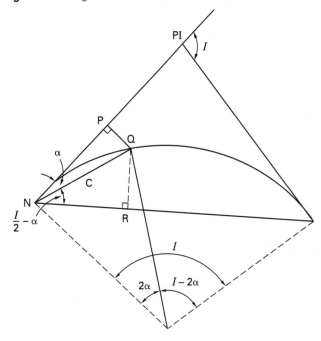

$$\text{NP} = \text{tangent distance} = \text{NQ} \cos \alpha \qquad \text{78.21}$$

$$\text{PQ} = \text{tangent offset} = \text{NQ} \sin \alpha \qquad \text{78.22}$$

$$\text{NQ} = C = 2R \sin \alpha \qquad \text{78.23}$$

$$\text{NP} = (2R \sin \alpha) \cos \alpha$$
$$= C \cos \alpha \qquad \text{78.24}$$

$$\text{PQ} = (2R \sin \alpha) \sin \alpha$$
$$= 2R \sin^2 \alpha \qquad \text{78.25}$$

7. CURVE LAYOUT BY CHORD OFFSET

$$\text{NR} = \text{chord distance} = \text{NQ} \cos \left(\frac{I}{2} - \alpha \right)$$

$$= (2R \sin \alpha) \cos \left(\frac{I}{2} - \alpha \right)$$

$$= C \cos \left(\frac{I}{2} - \alpha \right) \qquad \text{78.26}$$

$$\text{RQ} = \text{chord offset} = \text{NQ} \sin \left(\frac{I}{2} - \alpha \right)$$

$$= (2R \sin \alpha) \sin \left(\frac{I}{2} - \alpha \right)$$

$$= C \sin \left(\frac{I}{2} - \alpha \right) \qquad \text{78.27}$$

8. HORIZONTAL CURVES THROUGH POINTS

$$\alpha = \arctan \frac{y}{x} \qquad 78.28$$

$$m = \sqrt{x^2 + y^2} \qquad 78.29$$

$$\gamma = 90° - \frac{I}{2} - \alpha \qquad 78.30$$

$$\phi = 180° - \arcsin\left(\frac{\sin\gamma}{\cos\left(\dfrac{I}{2}\right)}\right) \qquad 78.31$$

$$\theta = 180° - \gamma - \phi \qquad 78.32$$

$$\frac{\sin\theta}{m} = \frac{\sin\phi\cos\dfrac{I}{2}}{R} \qquad 78.33$$

Figure 78.6 *Horizontal Curve Through a Point (I known)*

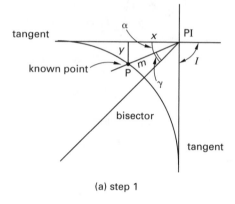

(a) step 1

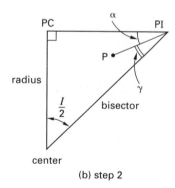

(b) step 2

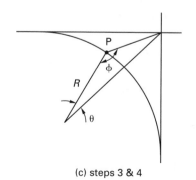

(c) steps 3 & 4

10. SUPERELEVATION

$$F_c = \frac{m v_t^2}{R} \quad \text{[consistent units]} \qquad 78.34$$

$$e = \tan\phi = \frac{v^2}{gR} \quad \text{[consistent units]} \qquad 78.35$$

$$e = \tan\phi = \frac{v^2}{gR} - f_s \quad \text{[consistent units]} \qquad 78.36$$

$$e = \tan\phi = \frac{v_{mph}^2}{15R} - f_s \qquad 78.37(b)$$

$$f_s = 0.16 - \frac{0.01(v_{mph} - 30)}{10} \quad [< 50 \text{ mph}] \qquad 78.38$$

$$f_s = 0.14 - \frac{0.02(v_{mph} - 50)}{10} \quad [50 \text{ to } 70 \text{ mph}] \qquad 78.39$$

11. TRANSITIONS TO SUPERELEVATION

$$T_R = \frac{wp}{\text{SRR}} \qquad 78.40$$

$$L = \frac{we}{\text{SRR}} \qquad 78.41$$

12. SUPERELEVATION OF RAILROAD LINES

$$E = \frac{G_{eff} v^2}{gR} \quad \text{[railroads]} \qquad 78.42$$

13. STOPPING SIGHT DISTANCE

$$S = \left(1.47 \frac{\frac{\text{ft}}{\text{sec}}}{\frac{\text{mi}}{\text{hr}}}\right) t_p v_{mph} + \frac{v_{mph}^2}{30(f + G)} \qquad 78.43(b)$$

15. MINIMUM HORIZONTAL CURVE LENGTH FOR STOPPING DISTANCE

$$S = \left(\frac{R}{28.65}\right)\left(\arccos\frac{R - \text{HSO}}{R}\right) \qquad 78.44$$

$$\text{HSO} = R(1 - \cos\theta) = R\left(1 - \cos\frac{DS}{200}\right)$$

$$= R\left(1 - \cos\frac{28.65 S}{R}\right) \qquad 78.45$$

16. VERTICAL CURVES

$$R = \frac{G_2 - G_1}{L} \quad \text{[may be negative]} \quad 78.46$$

$$\text{elev}_x = \frac{R}{2}x^2 + G_1 x + \text{elev}_{\text{BVC}} \quad 78.47$$

$$x_{\text{turning point}} = \frac{-G_1}{R} \quad \text{[in stations]} \quad 78.48$$

$$M_{\text{ft}} = \frac{AL_{\text{sta}}}{8} \quad 78.49$$

Figure 78.10 *Symmetrical Parabolic Vertical Curve*

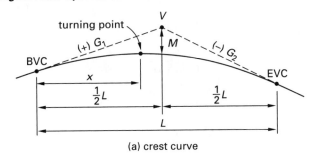

(a) crest curve

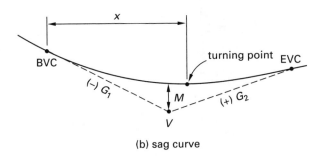

(b) sag curve

17. VERTICAL CURVES THROUGH POINTS

$$s = \sqrt{\frac{\text{elev}_{\text{E}} - \text{elev}_{\text{G}}}{\text{elev}_{\text{E}} - \text{elev}_{\text{F}}}} \quad 78.50$$

$$L = \frac{2d(s+1)}{s-1} \quad 78.51$$

Figure 78.11 *Vertical Curve with an Obstruction*

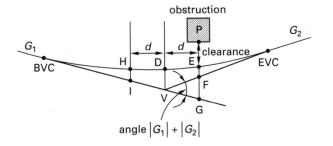

18. VERTICAL CURVE TO PASS THROUGH TURNING POINT

$$L = \frac{2(\text{elev}_{\text{V}} - \text{elev}_{\text{TP}})}{G_1 \left(\dfrac{G_1}{G_2 - G_1} + 1 \right)} \quad 78.52$$

19. MINIMUM VERTICAL CURVE LENGTH FOR SIGHT DISTANCES (CREST CURVES)

$$L = \frac{AS^2}{200 \left(\sqrt{h_1} + \sqrt{h_2} \right)^2} \quad [S < L] \quad 78.53$$

$$L = 2S - \frac{200 \left(\sqrt{h_1} + \sqrt{h_2} \right)^2}{A} \quad [S > L] \quad 78.54$$

$$L = \frac{AS^2}{800(C - 5)} \quad [S < L] \quad 78.55(b)$$

$$L = 2S - \frac{800(C - 5)}{A} \quad [S > L] \quad 78.56(b)$$

Table 78.4 *AASHTO Required Lengths of Curves on Grades*[a]

	stopping sight distance[b] (crest curves)	passing sight distance[c] (crest curves)	stopping sight distance (sag curves)
SI units			
$S < L$	$L = \dfrac{AS^2}{658}$	$L = \dfrac{AS^2}{864}$	$L = \dfrac{AS^2}{120 + 3.5S}$
$S > L$	$L = 2S - \dfrac{658}{A}$	$L = 2S - \dfrac{864}{A}$	$L = 2S - \dfrac{120 + 3.5S}{A}$
U.S. units			
$S < L$	$L = \dfrac{AS^2}{2158}$	$L = \dfrac{AS^2}{2800}$	$L = \dfrac{AS^2}{400 + 3.5S}$
$S > L$	$L = 2S - \dfrac{2158}{A}$	$L = 2S - \dfrac{2800}{A}$	$L = 2S - \dfrac{400 + 3.5S}{A}$

[a] $A = |G_2 - G_1|$, absolute value of the algebraic difference in grades, in percent.
[b] The driver's eye is 3.5 ft (1080 mm) above road surface, viewing an object 2.0 ft (600 mm) high.
[c] The driver's eye is 3.5 ft (1080 mm) above road surface, viewing an object 3.5 ft (1080 mm) high.

Compiled from *A Policy on Geometric Design of Highways and Streets*, Ch. 3, copyright © 2004 by the American Association of State Highway and Transportation Officials, Washington, D.C.

20. DESIGN OF CREST CURVES USING *K*-VALUE

$$K = \frac{L}{A} = \frac{L}{|G_2 - G_1|} \quad \text{[always positive]} \quad 78.57$$

22. MINIMUM VERTICAL CURVE LENGTH FOR COMFORT: SAG CURVES

$$L_{\text{ft}} = \frac{A\text{v}_{\text{mph}}^2}{46.5} \qquad \text{78.58(b)}$$

25. SPIRAL CURVES

Figure 78.17 *Spiral Curve Geometry*

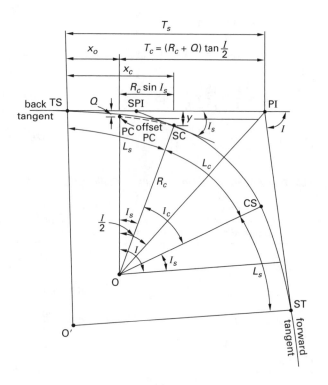

$$L_{s,\text{ft}} \approx \frac{1.6\text{v}_{\text{mph}}^3}{R_{\text{ft}}} \qquad \text{78.62(b)}$$

$$I_s = \left(\frac{L_s}{100}\right)\left(\frac{D}{2}\right) = \frac{L_s D}{200} \qquad \text{78.63}$$

$$I = I_c + 2I_s \qquad \text{78.64}$$

$$L = L_c + 2L_s \qquad \text{78.65}$$

$$\alpha_s = \tan^{-1}\frac{y}{x} \approx \frac{y}{x} \approx \frac{I_s}{3} \quad [I_s < 20°] \qquad \text{78.66}$$

$$\frac{\alpha_{\text{P}}}{\alpha_s} = \frac{I_{\text{P}}}{I_s} = \left(\frac{L_{\text{P}}}{L_s}\right)^2 \qquad \text{78.67}$$

$$y = x\tan\alpha_{\text{p}} \approx x\alpha_{\text{p,radians}}$$

$$= \frac{xI_{s,\text{radians}}}{3}$$

$$= \left(\frac{xL_s D_{\text{degrees}}}{(200)(3)}\right)\left(\frac{\pi}{180°}\right)$$

$$= \frac{xL_s}{6R_c} \quad [I_s < 20°] \qquad \text{78.68}$$

$$T_c = (R_c + Q)\tan\frac{I}{2} \qquad \text{78.69}$$

$$Q = y - R_c(1 - \cos I_s)$$

$$= \frac{L_s^2}{6R_c} - R_c(1 - \cos I_s) \qquad \text{78.70}$$

Construction

CERM Chapter 79
Construction Earthwork

Chapter, section, equation, figure, and table numbers correspond to CERM. For additional study material, go to the corresponding chapter and section number in CERM.

3. SWELL AND SHRINKAGE

$$V_l = \left(\frac{100\% + \% \text{ swell}}{100\%}\right) V_b = \frac{V_b}{L} \qquad 79.1$$

$$V_c = \left(\frac{100\% - \% \text{ shrinkage}}{100\%}\right) V_b \qquad 79.2$$

Summary of Excavation Soil Factors (bonus material)

quantity	symbol	example value or units	formulas		
swell factor; heaped factor [bank to loose]	SF	0.20	$\dfrac{\text{LCY}}{\text{BCY}} - 1$	$\dfrac{1}{\text{LF}} - 1$	$\dfrac{\gamma_{\text{bank}}}{\gamma_{\text{loose}}} - 1$
shrinkage factor; shrink factor [bank to compacted]	DF	0.15	$1 - \dfrac{\text{CCY}}{\text{BCY}}$		$1 - \dfrac{\gamma_{\text{bank}}}{\gamma_{\text{compacted}}}$
load factor [loose to bank]	LF	0.83	$\dfrac{\text{BCY}}{\text{LCY}}$	$\dfrac{1}{1 + \text{SF}}$	$\dfrac{\gamma_{\text{loose}}}{\gamma_{\text{bank}}}$
fill factor; dipper factor; fillability	FF	0.85	$\dfrac{\text{occupied volume}}{\text{solid capacity of bucket, dipper, or truck}}$		
operator factor	OF	0.93	$\dfrac{\text{actual production}}{\text{ideal production}}$		
loose volume; loose cubic yards	LCY	120 yd³	$(1 + \text{SF})\text{BCY}$	$\dfrac{\text{BCY}}{\text{LF}}$	$\dfrac{\gamma_{\text{bank}}}{\gamma_{\text{loose}}} \times \text{BCY}$
compacted volume; compacted cubic yards	CCY	85 yd³	$(1 - \text{DF})\text{BCY}$	$\left(\dfrac{1 - \text{DF}}{1 + \text{SF}}\right)\text{LCY}$	$\dfrac{\gamma_{\text{bank}}}{\gamma_{\text{compacted}}} \times \text{BCY}$
bank volume; bank cubic yards	BCY	100 yd³	$\dfrac{\text{LCY}}{1 + \text{SF}}$	$\dfrac{\text{CCY}}{1 - \text{DF}}$	
swell	SWL	20 yd³	LCY − BCY (increase)	SF × BCY	$\dfrac{1 - \text{LF}}{\text{LF}} \times \text{BCY}$
		20%	$\dfrac{\text{LCY} - \text{BCY}}{\text{BCY}} \times 100\%$ (increase)	SF × 100%	$\dfrac{1 - \text{LF}}{\text{LF}} \times 100\%$
shrinkage	SHR	15 yd³	BCY − CCY (decrease)	DF × BCY	
		15%	$\dfrac{\text{BCY} - \text{CCY}}{\text{BCY}} \times 100\%$ (decrease)	DF × 100%	

12. VOLUMES OF PILES

$$V = \left(\frac{h}{3}\right)\pi r^2 \quad \text{[cone]} \qquad 79.3$$

$$V = \left(\frac{h}{6}\right)b(2a + a_1) = \tfrac{1}{6}hb\left(3a - \frac{2h}{\tan\phi}\right)$$
$$\text{[wedge]} \qquad 79.4$$

$$V = \left(\frac{h}{6}\right)\left(ab + (a + a_1)(b + b_1) + a_1 b_1\right)$$

$$= \left(\frac{h}{6}\right)\left(ab + 4\left(a - \frac{h}{\tan\phi_1}\right)\left(b - \frac{h}{\tan\phi_2}\right)\right.$$
$$\left. + \left(a - \frac{2h}{\tan\phi_1}\right)\left(b - \frac{2h}{\tan\phi_2}\right)\right)$$
$$\begin{bmatrix}\text{frustrum of a}\\\text{rectangular pyramid}\end{bmatrix} \qquad 79.5$$

$$a_1 = a - \frac{2h}{\tan\phi_1} \qquad 79.6$$

$$b_1 = b - \frac{2h}{\tan\phi_2} \qquad 79.7$$

Figure 79.5 *Pile Shapes*

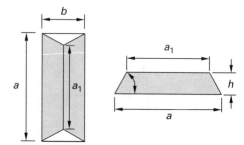

(a) cone

(b) wedge

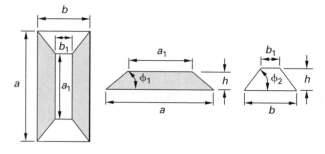

(c) frustrum of a rectangular pyramid

14. AVERAGE END AREA METHOD

$$V = \frac{L(A_1 + A_2)}{2} \qquad 79.8$$

$$V_{\text{pyramid}} = \frac{L A_{\text{base}}}{3} \qquad 79.9$$

15. PRISMOIDAL FORMULA METHOD

$$V = \left(\frac{L}{6}\right)(A_1 + 4A_m + A_2) \qquad 79.10$$

17. MASS DIAGRAMS

Figure 79.9 *Balance Line Between Two Points*

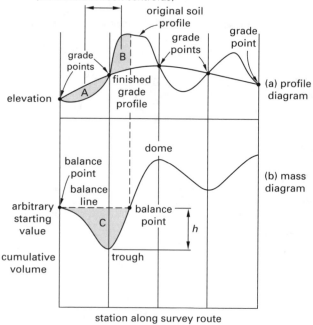

(a) profile diagram

(b) mass diagram

CERM Chapter 80
Construction Staking and Layout

Chapter, section, equation, figure, and table numbers correspond to CERM. For additional study material, go to the corresponding chapter and section number in CERM.

3. ESTABLISHING SLOPE STAKE MARKINGS

$$\text{HI} = \text{elev}_{\text{ground}} + \frac{\text{instrument height}}{\text{above the ground}}$$

$$= \text{elev}_{\text{ground}} + \text{ground rod} \qquad 80.1$$

$$\text{grade rod} = \text{HI} - \text{elev}_{\text{grade}} \qquad 80.2$$

$$h = \text{grade rod} - \text{ground rod} \qquad 80.3$$

$$d = \frac{w}{2} + hs \qquad 80.4$$

Table 80.1 *Common Stake Marking Abbreviations*

@	from the	INV	invert
¼SR	quarter point of slope rounding	ISS	intermediate slope stake
½SR	midpoint of slope rounding	JT	joint trench
ABUT	abutment	L	length or left
BC	begin curve	L/2	midpoint
BCH	bench	L/4	quarter point
BCR	begin curb return	LIP	lip
BEG	begin or beginning	L/O	line only
BK	back	LOL	lay-out line
BL	baseline	LT	left
BM	benchmark	MC	middle of curve
BR	bridge	MH	manhole
BSR	begin slope rounding	MP	midpoint
BSW	back of sidewalk	MSR	midpoint of slope rounding
BVC	begin vertical curve	OG	original ground
C	cut	O/S	offset
CL	centerline	PC	point of curvature
CB	catch basin	PCC	point of compound curve
CF	curb face	PG	pavement grade or profile grade
CGS	contour grading stake	PI	point of intersection
CONT	contour	POC	point on curve
CP	control point or catch point	POL	point on line
CR	curb return	POT	point on tangent
CS	curb stake	PP	power pole
CURB	curb	PPP	pavement plane projected
DAY	daylight	PRC	point of reverse curvature
DI	drop inlet or drainage inlet	PT	point of tangency
DIT	ditch	PVC	point of vertical curvature
DL	daylight	PVT	point of vertical tangency
DMH	drop manhole	QSR	quarterpoint of slope rounding
DS	drainage stake	R	radius
E	flow line	RGS	rough grade stake
EC	end of curve	RT	right
ECR	end of curb return	RP	radius point or reference point
EL	elevation	RPSS	reference point for slope stake
ELECT	electrical	ROW	right of way
ELEV	elevation	R/W	right of way
END	end of ending	SD	storm drain
EOP	edge of pavement	SE	superelevation
EP	edge of pavement	SHLD	shoulder
ES	edge of shoulder	SHO	shoulder
ESR	end slope grounding	SL	stationing line
ETW	edge of traveled way	SR	slope rounding
EVC	end of vertical curve	SS	sanitary sewer or slope stake
EW	end wall	STA	station
F	fill	STR	structure
FC	face of curb	SW	sidewalk
FDN	foundation	TBC	top back of curb
FE	fence	TBM	temporary benchmark
FG	finish grade	TC	toe of curb or top of curve
FGS	final grade stake	TOE	toe
FH	fire hydrant	TOP	top
FL	flow line	TP	turning point
FTG	footing	TW	traveled way
G	grade	VP	vent pipe
GRT	grate	WALL	wall
GTR	gutter	WL	water line
GUT	gutter	WM	water meter
HP	hinge point	WV	water valve
INL	inlet	WW	wingwall
INT	intersection		

Compiled from various sources. Not intended to be exhaustive. Differences may exist from agency to agency.

Figure 80.3 *Determining Stake Location*

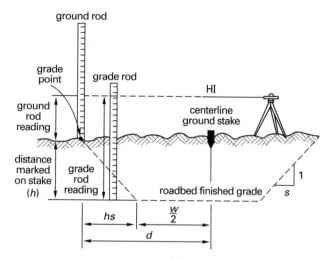

CERM Chapter 81
Building Codes and Materials Testing

Chapter, section, equation, figure, and table numbers correspond to CERM. For additional study material, go to the corresponding chapter and section number in CERM.

8. REQUIREMENTS BASED ON OCCUPANCY

See Table 81.2.

Table 81.2 *Occupancy Groups Summary*

occupancy	description	examples
A-1	assembly with fixed seats for viewing of performances or movies	movie theaters, live performance theaters
A-2	assembly for food and drink consumption	bars, restaurants, clubs
A-3	assembly for worship, recreation, etc., not classified elsewhere	libraries, art museums, conference rooms with more than 50 occupants
A-4	assembly for viewing of indoor sports	arenas
A-5	assembly for outdoor sports	stadiums
B	business for office or service transactions	offices, banks, educational above the 12th grade, post office
E	educational by > 5 people through 12th grade	grade schools, high schools, day cares with more than 5 children ages greater than 2.5 yr
F-1	factory moderate hazard	see code
F-2	factory low hazard	see code
H	hazardous—see code	see code
I-1	> 16 ambulatory people on 24 hr basis	assisted living, group home, convalescent facilities
I-2	medical care on 24 hr basis	hospitals, skilled care nursing
I-3	> 5 people restrained	jails, prisons, reformatories
I-4	daycare for > 5 adults or infants (< 2.5 yr)	daycare for infants
M	mercantile	department stores, markets, retail stores, drug stores, sales rooms
R-1	residential for transient lodging	hotels and motels
R-2	residential with 3 or more units	apartments, dormitories, condominiums, convents
R-3	1 or 2 dwelling units with attached uses or child care < 6, less than 24 hr care	bed and breakfast, small child care
R-4	residential assisted living where number of occupants > 5 but < 16	small assisted living
S	storage—see code	see code
U	utility—see code	see code
dwellings	must use International Residential Code	

Note: This is just a brief summary of the groups and examples of occupancy groups. Refer to the IBC for a complete list or check with local building officials when a use is not clearly stated or described in the code.

CERM Chapter 82
Construction and Jobsite Safety

Chapter, section, equation, figure, and table numbers correspond to CERM. For additional study material, go to the corresponding chapter and section number in CERM.

2. TRENCHING AND EXCAVATION

OSHA Soil Classifications (OSHA 1926.652 App. A)
(bonus material)

Type A soils are cohesive soils with an unconfined compressive strength of 1.5 tons per square foot (144 kPa) or greater. Examples of type A cohesive soils are clay, silty clay, sandy clay, clay loam, and in some cases, silty clay loam and sandy clay loam. (No soil is type A if it is fissured, is subject to vibration of any type, has previously been disturbed, is part of a sloped, layered system where the layers dip into the excavation on a slope of four horizontal to one vertical or greater, or has seeping water.)

Type B soils are cohesive soils with an unconfined compressive strength greater than 0.5 tons per square foot (48 kPa) but less than 1.5 (144 kPa). Examples of type B soils are angular gravel; silt; silt loam; previously disturbed soils unless otherwise classified as type C; soils that meet the unconfined compressive strength or cementation requirements of type A soils but are fissured or subject to vibration; dry unstable rock; and layered systems sloping into the trench at a slope less than four horizontal to one vertical (but only if the material would be classified as a type B soil).

Type C soils are cohesive soils with an unconfined compressive strength of 0.5 tons per square foot (48 kPa) or less. Type C soils include granular soils such as gravel, sand and loamy sand, submerged soil, soil from which water is freely seeping, and submerged rock that is not stable. Also included in this classification is material in a sloped, layered system where the layers dip into the excavation or have a slope of four horizontal to one vertical or greater.

Layered geological strata include soils that are configured in layers. Where a layered geologic structure exists, the soil must be classified on the basis of the soil classification of the weakest soil layer. Each layer may be classified individually if a more stable layer lies below a less stable layer (e.g., where a type C soil rests on top of stable rock).

Maximum Allowable Slopes (OSHA 1926.652 App. B)
(bonus material)

soil or rock type	maximum allowable slopes (H:V)[1] for excavations less than 20 ft deep[3]	
stable rock	vertical	(90°)
type A[2]	$^3/_4$:1	(53°)
type B	1:1	(45°)
type C	$1^1/_2$:1	(34°)

[1]Numbers shown in parentheses next to maximum allowable slopes are angles expressed in degrees from the horizontal. Angles have been rounded off.

[2]A short-term maximum allowable slope of $^1/_2$H:1V (63°) is allowed in excavations in type A soil that are 12 ft (3.67 m) or less in depth. Short-term maximum allowable slopes for excavations greater than 12 ft (3.67 m) in depth must be $^3/_4$H:1V (53°).

[3]Sloping or benching for excavations greater than 20 ft deep must be designed by a registered professional engineer.

Sloping and Benching (OSHA 1926.652 App. B)
(bonus material)

Excavations Made in Type A Soil

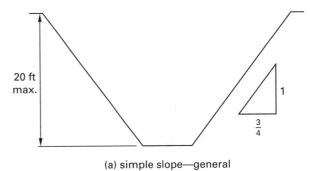

(a) simple slope—general

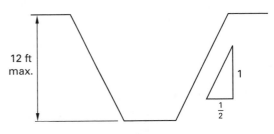

(b) simple slope—short term
(less than 24 hours)

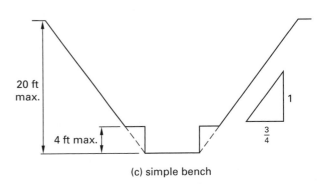

(c) simple bench

(continued)

Excavations Made in Type A Soil *(continued)*

Excavations Made in Type B Soil

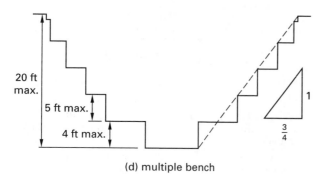

(d) multiple bench

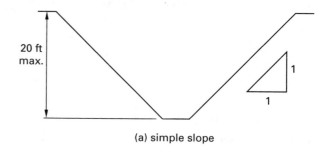

(a) simple slope

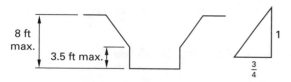

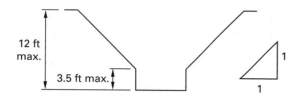

(e) unsupported vertical sided lower portion
(maximum 8 ft in depth)

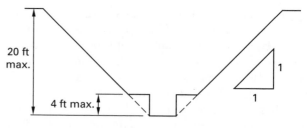

(b) single bench
(allowed in cohesive soil only)

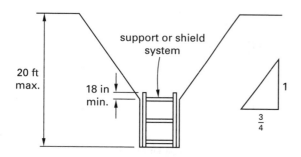

(f) unsupported vertical sided lower portion
(maximum 12 ft in depth)

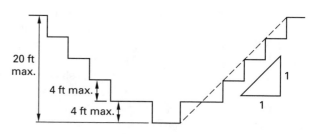

(c) multiple bench
(allowed in cohesive soil only)

(g) supported or shielded
vertical sided lower portion

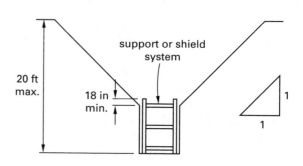

(d) supported or shielded
vertical sided lower portion

Excavations Made in Type C Soil

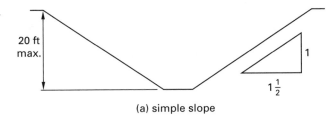

(a) simple slope

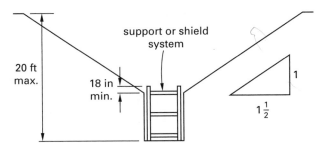

(b) supported or shielded
vertical sided lower portion

No stand-alone benching allowed in type C soil.

For sloping requirements in layered soils, refer to OSHA Sec. 1926 Subpart P App. B.

5. POWER LINE HAZARDS

$$\text{line clearance} = 10 \text{ ft} + (0.4 \text{ in})(V_{kV} - 50 \text{ kV}) \quad 82.1(b)$$

7. NOISE

Table 82.1 *Typical Permissible Noise Exposure Levels*[a]

sound level (dBA)	exposure (hr/day)
90	8
92	6
95	4
97	3
100	2
102	$1\frac{1}{2}$
105	1
110	$\frac{1}{2}$
115	$\frac{1}{4}$ or less

[a]without hearing protection
Source: OSHA Sec. 1910.95, Table G-16

Systems, Management, and Professional

CERM Chapter 85
Project Management, Budgeting, and Scheduling

> Chapter, section, equation, figure, and table numbers correspond to CERM. For additional study material, go to the corresponding chapter and section number in CERM.

4. SOLVING A CPM PROBLEM

ES	EF
LS	LF

key

ES: Earliest Start

EF: Earliest Finish

LS: Latest Start

LF: Latest Finish

step 1: Place the project start time or date in the **ES** and **EF** positions of the start activity. The start time is zero for relative calculations.

step 2: Consider any unmarked activity, all of whose predecessors have been marked in the **EF** and **ES** positions. (Go to step 4 if there are none.) Mark in its **ES** position the largest number marked in the **EF** position of those predecessors.

step 3: Add the activity time to the **ES** time and write this in the **EF** box. Go to step 2.

step 4: Place the value of the latest finish date in the **LS** and **LF** boxes of the finish mode.

step 5: Consider unmarked predecessors whose successors have all been marked. Their **LF** is the smallest **LS** of the successors. Go to step 7 if there are no unmarked predecessors.

step 6: The **LS** for the new node is **LF** minus its activity time. Go to step 5.

step 7: The slack time for each node is **LS−ES** or **LF−EF**.

step 8: The critical path encompasses nodes for which the slack time equals **LS−ES** from the start node. There may be more than one critical path.

6. STOCHASTIC CRITICAL PATH MODELS

$$\mu = \tfrac{1}{6}(t_{\text{minimum}} + 4t_{\text{most likely}} + t_{\text{maximum}}) \qquad 85.1$$

$$\sigma = \tfrac{1}{6}(t_{\text{maximum}} - t_{\text{minimum}}) \qquad 85.2$$

$$p\{\text{duration} > D\} = p\{t > z\} \qquad 85.3$$

$$z = \frac{D - \mu_{\text{critical path}}}{\sigma_{\text{critical path}}} \qquad 85.4$$

Planimeters *(bonus material)*

$$\text{area} = 2\pi r_{\text{wheel}} N L_{\text{wheel arm}}$$

CERM Chapter 86
Engineering Economic Analysis

> Chapter, section, equation, figure, and table numbers correspond to CERM. For additional study material, go to the corresponding chapter and section number in CERM.

11. SINGLE-PAYMENT EQUIVALENCE

$$F = P(1 + i)^n \qquad 86.2$$

$$P = F(1 + i)^{-n} = \frac{F}{(1 + i)^n} \qquad 86.3$$

12. STANDARD CASH FLOW FACTORS AND SYMBOLS

Table 86.1 *Discount Factors for Discrete Compounding*

factor name	converts	symbol	formula
single payment compound amount	P to F	$(F/P, i\%, n)$	$(1+i)^n$
single payment present worth	F to P	$(P/F, i\%, n)$	$(1+i)^{-n}$
uniform series sinking fund	F to A	$(A/F, i\%, n)$	$\dfrac{i}{(1+i)^n - 1}$
capital recovery	P to A	$(A/P, i\%, n)$	$\dfrac{i(1+i)^n}{(1+i)^n - 1}$
uniform series compound amount	A to F	$(F/A, i\%, n)$	$\dfrac{(1+i)^n - 1}{i}$
uniform series present worth	A to P	$(P/A, i\%, n)$	$\dfrac{(1+i)^n - 1}{i(1+i)^n}$
uniform gradient present worth	G to P	$(P/G, i\%, n)$	$\dfrac{(1+i)^n - 1}{i^2(1+i)^n} - \dfrac{n}{i(1+i)^n}$
uniform gradient future worth	G to F	$(F/G, i\%, n)$	$\dfrac{(1+i)^n - 1}{i^2} - \dfrac{n}{i}$
uniform gradient uniform series	G to A	$(A/G, i\%, n)$	$\dfrac{1}{i} - \dfrac{n}{(1+i)^n - 1}$

24. CHOICE OF ALTERNATIVES: COMPARING ONE ALTERNATIVE WITH ANOTHER ALTERNATIVE

Capitalized Cost Method

$$\text{capitalized cost} = \text{initial cost} + \frac{\text{annual costs}}{i} \qquad \text{86.19}$$

$$\text{capitalized cost} = \text{initial cost} + \frac{\text{EAA}}{i}$$

$$= \text{initial cost} + \frac{\text{present worth}}{\text{of all expenses}} \qquad \text{86.20}$$

25. CHOICE OF ALTERNATIVES: COMPARING AN ALTERNATIVE WITH A STANDARD

Benefit-Cost Ratio Method

$$B/C = \frac{\Delta^{\text{user}}_{\text{benefits}}}{\Delta^{\text{investment}}_{\text{cost}} + \Delta\text{maintenance} - \Delta^{\text{residual}}_{\text{value}}} \qquad \text{86.21}$$

26. RANKING MUTUALLY EXCLUSIVE MULTIPLE PROJECTS

$$\frac{B_2 - B_1}{C_2 - C_1} \geq 1 \quad [\text{alternative 2 superior}] \qquad \text{86.22}$$

36. DEPRECIATION METHODS

Straight Line Method

$$D = \frac{C - S_n}{n} \qquad 86.25$$

Sum-of-the-Years' Digits Method

$$T = \tfrac{1}{2}n(n + 1) \qquad 86.27$$

$$D_j = \frac{(C - S_n)(n - j + 1)}{T} \qquad 86.28$$

Double Declining Balance Method

$$D_{\text{first year}} = \frac{2C}{n} \qquad 86.29$$

$$D_j = \frac{2\left(C - \sum\limits_{m=1}^{j-1} D_m\right)}{n} \qquad 86.30$$

$$d = \frac{2}{n} \qquad 86.31$$

$$D_j = dC(1 - d)^{j-1} \qquad 86.32$$

Statutory Depreciation Systems

$$D_j = C \times \text{factor} \qquad 86.33$$

38. BOOK VALUE

$$\text{BV}_j = C - \sum_{m=1}^{j} D_m \qquad 86.43$$

41. BASIC INCOME TAX CONSIDERATIONS

$$t = s + f - sf \qquad 86.44$$

45. RATE AND PERIOD CHANGES

$$\phi = \frac{r}{k} \qquad 86.53$$

$$i = (1 + \phi)^k - 1$$

$$= \left(1 + \frac{r}{k}\right)^k - 1 \qquad 86.54$$

53. BREAK-EVEN ANALYSIS

$$C = f + aQ \qquad 86.58$$

$$R = pQ \qquad 86.59$$

$$Q^* = \frac{f}{p - a} \qquad 86.60$$

56. INFLATION

$$i' = i + e + ie \qquad 86.62$$

Index

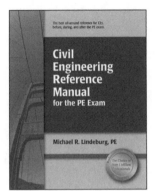

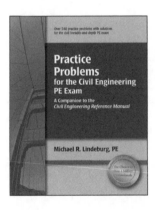

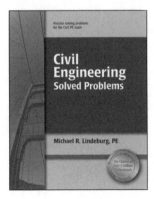